AVID

READER

PRESS

# GOLF'S HOLY WAR

## THE BATTLE FOR THE SOUL OF A
## GAME IN AN AGE OF SCIENCE

# BRETT CYRGALIS

**AVID READER PRESS**

NEW YORK   LONDON   TORONTO   SYDNEY   NEW DELHI

AVID READER PRESS
An Imprint of Simon & Schuster, Inc.
1230 Avenue of the Americas
New York, NY 10020

First Avid Reader Press hardcover edition May 2020

AVID READER PRESS and colophon are trademarks of Simon & Schuster, Inc.

For information about special discounts for bulk purchases,
please contact Simon & Schuster Special Sales
at 1-866-506-1949 or business@simonandschuster.com.

The Simon & Schuster Speakers Bureau can bring authors to your live event.
For more information or to book an event, contact the
Simon & Schuster Speakers Bureau at 1-866-248-3049
or visit our website at www.simonspeakers.com.

Interior design by Ruth Lee-Mui

Manufactured in the United States of America

1   3   5   7   9   10   8   6   4   2

Library of Congress Cataloging-in-Publication Data is available.

ISBN 978-1-4767-0759-4
ISBN 978-1-4767-0761-7 (ebook)

*To Claire*

The most beautiful experience we can have is the mysterious. It is the fundamental emotion which stands at the cradle of true art and true science.

—Albert Einstein, 1931

# CONTENTS

PROLOGUE

# NEW JERSEY

Put your hand high behind your head. Wiggle your fingers. Now look up and see how close your hand is to where you thought it was.

Knowing the whereabouts of your body parts is called proprioception, and we're all different levels of bad at it. Maybe your proprioception is better than your neighbor's, but when you get into positions that you are not familiar with, you will lose a large sense of where you are.

I learned this phenomenon from Skip Latella, who shocked me with what he said while giving a presentation at a golf outing. Latella was supporting a female golfer standing on hard rubber balloons called Flexor discs. As she stood there, he helped her back into a motion like the golf swing, and he said something that sounded entirely out of place:

"See, standing on the discs creates static electricity. That static electricity opens the neurotransmitters in your brain. Coupled with the innervating of deep muscle tissue, the communication between the body and brain is more open, and you're relearning a complex motor function." He paused and raised his hands. "You're teaching your brain and body subconsciously."

Because proprioception is so flawed, consciously telling your body what to do in any precise manner is a flawed way of changing motor patterns. Try to take a golf club back and then explain where your hands are. You know what your swing *feels* like to you, but that's almost never the reality of the situation, especially during a quick and complicated motion. If you know what Tiger Woods's swing looks like—for the sake of this argument, let's say his swing circa 2000, the specifics and timelines of which we will get into with much detail later—why can't you just do that? It's because you have no idea how to tell your body how to do that. It's like trying to tell a calculator to launch a rocket.

"My goal was to learn how the body moves in relation to its neurophysiological limitations," Latella said later. "I'm just trying to max out everyone's potential."

By doing that, Latella was trying to bypass the conscious mind. Yet wasn't the entire history of golf instruction—really, the entire history of education—based on the conscious mind? Didn't all teachers give verbal directions to their students, then ask the students to execute those directions?

*Is this where science has taken the game?*

"Are there any big-name teachers that use this?" I asked.

Latella stopped and thought for a minute.

"You know who gets it?" he finally said. "David Glenz."

I called Glenz at his academy, which at the time was located within a sprawling place in the Kittatinny Mountains of northwestern New

Jersey called Black Bear Golf Club, part of the Crystal Springs Resort. His voice was raspy and soft, with a tinge of what I would come to learn was Oregonian twang. He had taught some great players in his time, evidenced by *Golf Digest* naming him National Teacher of the Year in 1998.

I told him about my conversation with Latella, and Glenz listened intently. I said I would love to watch him teach with the Flexor discs, then sit down and talk about how golf instruction has changed.

"Well," Glenz said, "there's a conflict when you teach golf trying to use conscious information."

I didn't know a teacher could say something like that. Didn't that statement undercut an entire industry?

When I arrived at the academy a few days later, it was a sweltering-hot day in June, with little breeze and few clouds. It smelled wonderfully of earth. Glenz was there with his student Mike DeFazio. He was thirty-seven years old, in good shape, and carried a seven handicap; he could probably have been better if he hadn't had back surgery six months prior. He said he used to be a pro bowler, but now worked as a contractor.

DeFazio was balancing himself on the Flexor discs, and his clubs were scattered behind him on the ground. His first three months of lessons were just like this, but Glenz never explained the science behind the discs. After DeFazio stepped off, he hit a midiron pure and straight. Turning to me, he said that the discs weren't there to help people repeat the swing, but "to help with muscle memory and balance."

Later I told Glenz about his student's self-conflicting response. Glenz smirked and then lit a Marlboro Red. As he thought, he let the smoke exit through his nostrils.

"The trick is—that's the real teaching part," Glenz said. "Anybody can look at video and say, 'You're here, you should be there.' But how do you make your body change in order to have the golf ball perform?"

Then I walked about a hundred paces down the range and met Henry Ellison, which is when things started to get really weird.

When I first saw Ellison, he was teaching golf six inches from a student's face. Later, when I was the student, it was impossible not to notice the graying of his five-o'clock shadow and the white lines around his lips that said this man needed a drink of water an hour ago—but the thought hadn't even crossed his mind.

He spoke in platitudes, in self-made proverbs, and the passion with which he conveyed these ideas was infectious. His voice would rise in pitch and volume at the end of every statement, plumbing for a connection.

"Do you understand?" he would say often. "Do you *really understand*?"

On that first day, Ellison was talking to Robert Adelman, fifty-six, from the wealthy suburb of Scarsdale, New York. He had begun playing with his teenage daughter, who quickly grew better than him and quickly grew embarrassed by how bad her dad was.

Frustrated, Adelman was on the verge of quitting the game. But just for the sake of thoroughness, he figured he would give one last teacher a try before giving it up for good. His close friend Dan had taken lessons from Ellison for years—"as long as I can remember," said Dan, an aloof single-digit-handicapper with a strange and distant smile. Dan was there just to watch from under his wide-brimmed straw hat, not giving a last name or hometown or any sense of his feet being on the ground. I should've realized then that I was treading into strange waters.

Before any student was even allowed to pick up a club, Ellison began what he called "detox." Through a seemingly never-ending stream of questions, he started to extract all the golf-swing rhetoric that amasses inside anyone who has ever tried to hit a golf ball with conviction. With the proliferation of high-end technology such as ball-flight monitors and 3-D motion analysis—again, which we will get into with great detail later—Ellison thought the vast majority of teachers believed they had found a tangible way to "prove" their theories about the golf swing. Either through direct lessons or magazines or television shows, that information leaked into the general psyche of the golfing public.

"Everyone wants the game to be more sure, more constant, the same all the time," Ellison said. "It's almost like wanting it to be the same all the time rubs up against the beauty of it. There are certain forces that are nature-bound that can't be captured. It's mysterious."

As I would learn, Ellison had fallen deeply in love with science and technology as a budding player. He grew up in Newton, Massachusetts, the son of a mostly absentee alcoholic father. He had a younger sister, and while his mother worked, Henry was the adult. He played a lot of sports and thought golf was boring. After graduating from Bowdoin College in Maine, Ellison got a job in the mid-1980s with a major financial firm in Manhattan, first working in commercial banking, then selling bonds. He eventually joined the revered Baltusrol Golf Club in New Jersey and became a competitive top-end amateur. Upon reaching this level, Ellison then left his job, left his condo in the suburbs, left his longtime girlfriend, and drove to Florida to play the mini-tours.

He had picked the brains of some of the best teachers in the country, including David Leadbetter, Hank Haney, Jimmy Ballard, and even Glenz. Ellison once went to an outing and experienced a spiritual connection with famed Canadian ball-striker Moe Norman. He said he videotaped the conversation, but I never got a chance to see it. His camera was omnipresent at that time, he said, as he would tape every swing he made—every one—and rush over to play it back so a screen could tell him what he just experienced.

But now, in front of Adelman, the only aid Ellison had was a single stick used for alignment. His divorce from video and technology had left a scar, and his impatience with new age teaching was clear. The "detox" continued, Adelman mentioning a swing tip and Ellison cutting him off midway through: "Anything else?"

After an alarmingly long time spent weeding out his student's preconceptions, Ellison asked Adelman for his personal swing thoughts.

"Well, I don't have any," Adelman said with a shrug. "When I get over the ball, I just want it all to be over. I'm so frustrated I just don't care. I just want to be done with it."

Ellison's eyes pinched at the sides and his head tilted, as if this was a genuinely sad statement for him to hear. But the sadness quickly gave way to excitement about another opportunity, another person he could lead out of the darkness and into the light.

"The more science investigates, the more spiritual it becomes," Ellison said. "Enjoy and learn is my attitude, and then you're not going to be miserable."

After forty-five minutes, Ellison had his student brushing an iron against the ground and pretending that the point where the driving-range grass went from fairway height to rough was where the ball was positioned. "All the stuff you're learning," Ellison said, "is just to get this without thinking about it."

"Why?"

"Because you're in the truth now!"

With that, this crazy-eyed teacher jumped backward in the air with his arms above his head and a small smile turning up the corners of his mouth. His thick cotton shirt was almost completely pulled out from his lightweight slacks, and he seemed not to notice—or care.

"Today was the best lesson I've ever gotten," Adelman said afterward. "Everything he said just seemed so intuitive."

Walking off the driving range, thirty-five minutes past the allotted time, Ellison was still talking, still asking his student if he understands— if he *really understands*. Adelman answered yes, not as an empty appeasement but as an affirmation of the serious philosophical connection that both men felt, even if neither could fully explain it.

"I want people to be self-realized, self-actualized, self-strengthening," Ellison said. "I want them to swing and I want it to be truthful, and I want them to be truthful."

He paused for a minute and looked up at a tall pine in the distance. Putting his right palm to his cheek, he slowly shook his head.

"I just give people what I wanted," he said.

When I caught up with Ellison again, he had left the Glenz academy. He departed on good terms, but the split seemed inevitable. He picked up work at a nearby driving range owned by a friend, a run-down place always about to be refurbished. There was a parking lot of loose gravel, a sign that read NEW CARPET—MINI GOLF OPEN, and a big vegetable farm across the street. It was late summer, and it was hot, and when the wind kicked up some dust, it felt as if it could have been another time and place. Ellison taught from the loosest definition of grass, aiming at telephone poles in the distance. His students followed. One by one, I listened to them tell different versions of the same wistful story.

When Ellison would give lessons, I'd watch. In the years since, I have never seen another teacher make such deep connections with his or her students. Afterward we would pull up two dirtied white plastic chairs, and for hours I would ask him questions, and he'd answer. He always wanted to make sure I understood what he was saying—that I *really understood*. He'd be holding a sand wedge, picking away at a clump of grass, looking down.

Then he'd look up for emphasis, and empathy. "I don't want to teach technology," he said. "I want to teach humanity. *Get it?*"

He told me about how he had had Lyme disease for almost thirty years without knowing it. The tick-borne disease, difficult to diagnose, left him fatigued all the time. He went to doctor after doctor and no one had an answer. It was just like taking golf lessons—all thought they had the answer, and no one did. He would occasionally be laid up in bed for days, thinking he was dying. He would have hallucinations. He would wake up in places and not remember how he got there.

He brought up an Einstein quote. I brought up relativity, and he cracked a small smile. "I've always had real trouble with time," he said. "I have trouble being *here*."

It all went back to a hiking trail through the woods when Ellison was twelve, where he was bitten by a tick. By the time he was older and pursuing his playing career, the disease had done its damage on his joints, and his back was constantly aching. It ached as he spoke to me.

The conversations I had with Henry were quiet and intimate. The game had touched him on some inner level, and he was letting me know about it. He made me feel closer to something so abstract; something that was changing every day, but the root of which was unchangeable.

To get a better understanding of this weird and wonderful game, to picture the battle waged between art and science, I started with two guys in New Jersey.

Golf is only one small example of how such exponential progress in the realm of science and technology has changed our modern world. It has entirely altered the way people think, and the way people approach reality. As the amount of information grows, the natural longing is to explain everything that was once thought unknowable. As science marches on, the feeling that it's possible to explicitly explain everything increases.

But defining every minute detail of the physical world rubs up against some inherent mysteries that have thus far described the human experience. The gap between those mysteries and our understanding closes slightly with each additional piece of information. But many people find that gap in human knowledge to be the place where beauty resides, where God resides. Many people find ambiguity to be the home of art.

So this war between art and science rages on, with golf as an apt example of the huge shift in the modern way of thinking. The opposing factions, each with their own basis of beliefs and ardent followers, battle for authority. There might be a middle ground, but the fissure is getting wider.

I've watched this divide first crack and then deepen. By seeking out some of the most influential people in the game's history, I've discovered a past that is ripe with clues about exactly what happened, why it happened, and where it might all go in the future. Like everything else in the rapidly shifting landscape of modernity, golf is in the midst of a monumental change. Not everybody is on board.

I wanted to discuss all this with Ellison, but I stopped being able to reach him. He was gone. Picked up and left for Florida, maybe. Maybe he was ill again. Maybe worse. Maybe I would hear from him again when I least expected it.

Or maybe Henry never existed at all. As in so many other corners of this strange and fascinating world, maybe it's those who are gone who teach you the most important lessons. And maybe chasing mysteries is the only way to move closer to truth, even if you never fully get there.

# ONE

# THE TWO CENTRAL TEXTS

Two books are paramount to understanding the state of modern golf, and most golfers haven't heard of either.

The first is titled *The Golfing Machine*, self-published in Seattle in 1969 by a part-time electrical engineer named Homer Kelley. The first edition was a tightly wound, 156-page instructional manual. Overly dense, incredibly technical, it was regarded as either impenetrable or essential.

The second book, titled *Golf in the Kingdom*, was published by the well-known Viking Press in 1972. Authored by Michael Murphy, the son of a wealthy California doctor, it was a loosely compiled sequence of semifictional experiences, some on a golf course, some not. Philosophically engaging, strangely mystical, it was either disregarded as lunacy or held dear to the heart.

Given that golf has the most extensive and eclectic literature of any sport, books are the natural place to look to understand the game.

There are epic histories, countless biographies (especially of Ben Hogan), and new instructional works that have continued to multiply over the past hundred years. So it is only logical that as science began to advance during the mid-twentieth century at a rate exponentially faster than at any other time in human history, this subset of literature advanced in turn.

Yet in the game of golf, as in the rest of the world, advancements in science and technology began to carve out a divide between the past, filled with ambiguities, and the projection of the future, filled with seemingly provable scientific certainties. As is often the case in industry upheaval, the majority of golfers fell in line with what was new. There was (and remains) a secular attack on previous psychological ideologies and technical methodologies, disregarded as the products of misled faith. The proponents of the new age find comfort in the cold exactness of *The Golfing Machine*. The backlash came from a much smaller group of people (albeit including some powerful figures) holding firm to their belief in what philosophy came before them, whether tangibly provable or not. Those proponents find truth in the strangeness of *Golf in the Kingdom*.

This Venn diagram has a middle ground, but one cannot love both of these works equally. The affinity one feels for one book or the other represents a fundamental decision each individual makes regarding the interpretation of reality, either as something solvable or something inherently mysterious.

This clash is a product of the modern age, and it happens that golf is the medium.

The 1982 Open Championship was to be Bobby Clampett's first trip to play professional golf overseas. At twenty-two years old, he was one of the game's rising stars, seemingly destined for greatness. In his first year at Brigham Young University in 1977 (although he wasn't a Mormon, he liked how nice the people were), Clampett was named Freshman of the Year as well as first-team All-America. That summer, he won both the prestigious Western Amateur and the Porter Cup, the latter coming off a second-round 62. After every victory, he would credit *The Golfing Machine*, and people would look at him sideways. He lost to John Cook in the semifinals of the 1978 U.S. Amateur, and soon thereafter *Golf World* magazine had had enough of making just short mentions of the book and ran a cover story titled "Bobby Clampett and *The Golfing Machine*."

"It's the bible of golf," Clampett told the magazine. "It's nothing to laugh at."

Nobody laughed as Clampett finished up his amateur career with three straight All-America honors and two consecutive Fred Haskins Awards as the best player in college. Before the final round of the 1980 NCAA tournament, Clampett stayed up most of the night with agent Hughes Norton from IMG, the leading sports talent agency, plotting his moneymaking future as a professional. He then shot the first 80 of his college career, and his BYU team came in second. That summer, he left behind amateurism and jumped into the PGA Tour, making six of ten cuts, including a top five at the Buick-Goodwrench Open, getting him playing privileges for the 1981 season.

Clampett's first full year was good if unspectacular, and despite not getting a win, he earned $184,710 to finish fourteenth on the money list, two spots ahead of a forty-one-year-old Jack Nicklaus. When 1982 started, Clampett came out with a handful of top fives, highlighted by a tie for third at the U.S. Open at his hometown course, Pebble Beach, where Tom Watson chipped in on the penultimate hole for one of the most memorable major victories of all time. Later that summer, it was time to head to Royal Troon, on the southwestern coast of Scotland, where the 111th Open Championship would be played. Clampett was one of the most interesting players to watch.

Thursday's opening round saw rain and cold and wind, and

Clampett wore traditional Scottish garb: a pair of white plus 2s, dark argyle knee socks, and a tam-o'-shanter. He shot a 5-under 67 and led Watson by two. The next day, in the warmth of the sunshine, he followed it up with a course-record 66, his two-day total of 11 under leading Nick Price by five shots.

That night, Clampett picked up the phone and called his teacher back in Carmel Valley in Northern California. Ben Doyle was the first "authorized instructor" of *The Golfing Machine*, having been introduced to the book when Homer Kelley walked into Doyle's pro shop at Broadmoor Golf Club in Seattle with a first edition back in 1969. The two had a long and tumultuous relationship revolving around their belief in Christian Science and this dense scientific text. Its message resonated deeply with Clampett and pushed him to the height of leading the Open.

"I might run away with this tournament," Clampett told his teacher. "I've done it before."

On the fifth hole the next day, Clampett chipped in for a birdie 2. He now led the oldest and most revered golf tournament in the world by seven shots with thirty-one holes to play. Stepping up to the sixth tee like a man on top of the world, he pounded a drive down the fairway, but seemingly out of nowhere a tight little pot bunker swallowed it up. He tried to chip out, hit the lip, and the ball went into another bunker. His next chip-out attempt hit another lip and just got out to the fairway. He then hit a metal wood way left into the crowd. He chipped from the rough into another bunker, right up against the sod-stacked face. He popped it out onto the green, twenty feet from the hole. He putted once, twice, and then dropped his head to write 8 on the scorecard.

Walking off the green, he looked back at the tract of land that rolled like waves in a calm ocean. No bunkers could be seen, and the sun was high in the sky, as if Andrew Wyeth had painted a Scottish countryside. But that ground was subtly violent, and it left Clampett forever scarred.

Summoning all his defiance, he then opened his mouth and stuck out his tongue.

The salty air was as bitter as it was sweet. The game's innate ability to push at just the right time was indelible, and it had pushed Clampett hard. No longer was it the laws of physics that were being tested, but something deeper, something along the ambiguous lines of character and mettle. There was no book to explain where Clampett was going, no guide with laws to lead him. Now, Clampett looked into the abyss of mystery. Now, with no guardrails, he fell.

He finished his third round with a 78, and even with a pair of gray plus 2s for Sunday, he shot a final-round 77. His four-round total of even par left him four shots behind the winner, Watson.

"What is more important in golf: character or technique?" Watson asked after having won his fourth of five Opens, just one short of Harry Vardon's record. "Character," he continued. "You have to have the guts to fight it out. There are days when you go out there and know you have the worst end of the deal, but the great players keep fighting.

"I feel very sorry for Bobby," Watson continued. "He may be crying right now, but I've cried before, and he'll learn to be tough."

Clampett struggled to accept the loss for a long time. Immediately after the tournament, his explanation of what had happened went back to concreteness of the book, and of geometry. "I had some compensating moves going on," he said. "Essentially, I was too steep coming into the ball. My swing relied too much on timing."

It was the beginning of the end for one of golf's most promising careers. Clampett did manage to win his first (and only) PGA Tour event later that fall, when most of the world's best players were on vacation. Wearing the same gray plus 2s he wore for the final round at Troon, he shot a final-round 64 to win the Southern Open in Columbus, Georgia. Yet in 1983, he made only half of his cuts and had no top tens. In 1984, he made sixteen of twenty-nine cuts, but again, not a single top ten.

Standing on sixth tee at Troon in 1982 was as close to the summit as Clampett would ever get.

Homer Kelley was a loner, his family moving from Clayton, Kansas, to Minneapolis in 1912 when he was five. After two years of miscellaneous and unfocused study at college, he was living in Tacoma, Washington, working as a cook at a billiard hall. His boss, James Cooksie, invited Kelley out to play his first round of golf on January 31, 1939. Battling through nerves, Kelley shot 116 on the 5,894-yard Meadow Park Golf Course, losing to the boss by just one.

"I hit the ball so well at the driving range," Kelley said, according to Scott Gummer's book *Homer Kelley's Golfing Machine*. "Why couldn't I do it on the course?"

Kelley didn't play again for another six months, teeing it up again with the boss in July of 1939 at the Highland Golf Course in Tacoma. He shot 77.

From there on, Kelley went to numerous local pros to find out how he had improved so much with no actual practice, and he began keeping diligent notes. He got descriptions of what to do, not explanations of why or how to do it. When he pushed for specific answers, he hit nothing but dead ends. According to Gummer's book, in the mid-1940s Byron Nelson came to Seattle for a golf clinic, and Homer got to ask the best player in the world at the time a question.

"I swing fast, though I don't want to," Kelley said. "I try to swing slowly, but there is no way. What causes that, and how can I stop it?"

Nelson was taken aback. "Gee," he said, "swing slow, I guess."

In 1941, Kelley got a job that paid sixty-two and a half cents an hour at Boeing Airplane Company in Seattle doing wiring for the B-17F bomber. He developed books of circuit diagrams and was soon transferred to the functional testing of the plane. Through the next decade, including the years of World War II, he missed significant portions of time at work due to injury, layoffs, and a strike. It gave him some free time, and his notes about golf grew into an idea to write a definitive instructional book. As if figuring out a circuit, Kelley would figure out the golf swing.

Despite being "dismissed as incompetent" from Boeing, in 1950 Kelley was hired at the Sand Point Naval Air Station just north of Seattle, jutting out into Lake Washington. He was on his third wife with no children and few hobbies besides Christian Science and his dog named Shadow. He rarely played golf, but he was obsessed with a physical explanation of the golf swing. Over the past twenty-eight years since he had first picked up the game, he had compiled a staggering amount of information. He then began to whittle it away to what he thought was an understandable format for a single book. He met a woman named Diane Chase one afternoon at a local driving range, and he used her as a model for some poorly lit black-and-white photographs in his garage studio to accompany his technical text.

The premise of the book is that if the golf swing can be described in the terms of physics and geometry, then it can be understood and ultimately controlled. "The relationships in the Golf Stroke can be explained scientifically only by geometry, because geometry is the science of relationships," Kelley wrote in the introduction, with capitalization used throughout as in a reference book, which is what he considered his work. "So learn Feel from Mechanics rather than Mechanics from Feel."

Homer Kelley does not propose a singular way to swing the club, but rather millions of variations depending on what best suits each person. It starts with "two basic Strokes—Hitting and Swinging. The geometry (for 'uncompensated' Strokes) is the same for both. . . . But, basically, the Physics of Hitting is Muscular Thrust, and of Swinging, Centrifugal Force. Hitting and Swinging seem equally efficient. The difference is in the players. If strong—Hit. If quick—Swing. If both— do either. Or both."

It is clear right away that *The Golfing Machine* is not an instructional book that claims to offer some newfound answer or to reveal any "secrets" of the game, the two basic premises of so many other works in that genre. Instead, Kelley explains detailed physical motions, and how they might be combined into a useful golf swing. It is a working guide for each player to discover and put together his or her own answer.

The basis of the assembly line to build this repetitive "machine" is for the player to establish a Stroke Pattern. To start that process, in the preface Kelley recommends reading the book out of sequence. First, read one list of chapter subsections—while "ignoring cross-reference numbers"—and once "you grasp the essentials (more or less)," you move on to the second list. Once that is completed, you begin "the preliminary assembly of your selected Pattern from Chapter 12."

The motion of the golf swing was broken down into 12 Sections (chapter 8): Preliminary Address (8-1), Impact Fix (8-2), Adjusted Address (8-3), Start Up (8-4), Backstroke (8-5), Top (8-6), Start Down (8-7), Down Stroke (8-8), Release (8-9), Impact (8-10), End of Follow Through (8-11), and Finish (8-12). That is then broken down into Three Zones (chapter 9) "of the action that is occurring throughout the 12 Sections listed in Chapter 8." Zone One is Body Control, which includes Pivot, Body, and Balance; Zone Two is Club Control, which includes Power, Arms, and Force; and Zone Three is Ball Control, which includes Purpose, Hands, and Direction.

Incorporated inside these 12 Sections and Three Zones are the 24 Basic Components (chapter 7) of the golf swing, which are the foundation of any Stroke Pattern. They are Grip—Basic; Grip—Type; Stroke—Basic; Stroke—Variation; Plane Line; Plane Angle—Basic; Plane Angle—Variation; Fix; Address; Hinge Action; Pressure Point Combination; Pivot; Shoulder Turn; Hip Turn; Hip Action; Foot Action; Left Wrist Action; Lag Loading; Trigger Type; Power Package

Assembly Point; Power Package Loading Action; Power Package Delivery Path; and Pack Package Release.

Now, a Catalog of Basic Component Variations (chapter 10) begins to complicate things. Here each of the 24 Basic Components is broken down into many, many options, and Kelley wrote that anything not included was omitted for a reason. In the first component (Grip—Basic, further defined as "Hand to Hand") he establishes five variations, briefly described in the text and shown in photos: Typical Overlapping Grip, Baseball Grip, Reverse Overlap, Interlocking, and Cross Hand Grip. In the second component (Grip—Type, further described as "Hands to Plane") he has seven variations, ones that also cross-reference Rotational Wrist Conditions (chapter 4-C) of the Left Wrist, Right Wrist, and the No. 3 Pressure Point.

Overall, there are 144 variations of the twenty-four components that constitute the twelve sections of the golf swing, parsed into three zones. The golfer then picks any combination of these variations to make up his or her own Stroke Pattern, which brings about the possibility of millions of combinations, some working better than others. Each component variation then needs to be monitored throughout the twelve sections and three zones, which would emphasize the necessity of an "authorized instructor," such as Ben Doyle.

Chapter 11 is a summary of all the variations, subtitled "Golf as a shopping list." Chapter 12 then begins giving examples of possible usable Stroke Patterns—as in, which variations of the twenty-four components go well together. Chapter 13 is then a warning list of "Non-Interchangeable Components," subtitled, "Golf as square pegs and round pegs."

Kelley wrote, "An important point for careful consideration—is the player benefited by this fragmentation of the Golf Stroke? Undoubtedly. Not only eventually, but immediately."

Given this staggering amount of technical rhetoric, most followers will lean back on two fundamental principles that are oversimplifications but became ubiquitous when discussing the book. First is a Flat Left Wrist at impact (for a right-handed player), meaning the wrist and the forearm are in alignment when the ball is struck. (Of course, the many variations depend on the Grip—Basic and Grip—Type chosen, among other differentiations among the twenty-four components. But, a flat left wrist is paramount.) The second principle is Sustain the Lag, which has become a calling card of sorts, almost a secret code spoken to identify those in the know. "Lag," Kelley wrote, "defines the condition of 'trailing,' or 'following,' and can, and usually should, exist to some degree at every point in the Stroke from feet to Clubhead." The idea of "sustaining" the lag helps to accrue power throughout the sections of the swing, as "every Lagging Component places a Drag on its preceding Component, which is proportional to the Rate of Acceleration of the leading component." This principle is so important that Clampett once did a science project on clubhead lag while en route to graduating Robert Louis Stevenson High School, in Pebble Beach, in three years.

Understandably, people of a certain scientific inclination would be drawn to such a detailed technical book and be so fervent in their adoration. For the first and only time, the golf swing was broken down to its elemental form. It was like J. J. Thomson discovering the electron in 1897.

But at a time when so many people longed for more science and more tangible evidence of the surrounding world, this technical instructional book with all the answers barely got off the ground. It remained the focus of only a small group of people, stemming from those who were likely in personal contact with Kelley. Like every instructional device, it was judged by its results, and critics swiftly attacked the book's teachings when Bobby Clampett fell from grace. It didn't help the book's reputation when the next best disciple player to come along, Mac O'Grady, notched two PGA Tour wins in the late 1980s before veering off into wild eccentricity rather than more success. At its peak, the book never sold more than a couple thousand copies a year. Kelley put out six editions (and even toned down some ideas for more commercial features in magazines, which alienated some of his most ardent followers) before he died of a heart attack while giving a presentation to the Georgia PGA on Valentine's Day in 1983.

Yet the book has had its biggest impact on some of the world's foremost golf instructors, who didn't flock to become authorized instructors (likely for fear of being pigeonholed), but found the information valuable. David Leadbetter, who taught many stars, including Nick Faldo at his height as the best player in the world in the 1990s, made all of the instructors at his schools read the book as background. Michael Hebron, the 1991 National PGA Teacher of the Year, who dramatically shifted gears into simpler learning environments later in his career, said he wouldn't feel comfortable in front of students without knowing the information in the book. Steve Elkington, with ten wins on the PGA Tour, including the 1995 PGA Championship, often practiced without hitting balls, using only the "Facts and Illusions" mat that Doyle created as a supplementary tool to the book.

In December 2002, two teaching professionals, one from Oregon named Joe Daniels and one from Georgia named Danny Eakins, bought from Homer's widow, Sally, the copyright to *The Golfing Machine*, its operations, Kelley's memorabilia, and the manuscript for an unpublished seventh edition. Daniels and Eakins were instrumental in helping Scott Gummer write his 2009 biography about Kelley and his work.

Any growing attention was amplified in 2015 when Bryson DeChambeau (of Southern Methodist University) became just the fifth player in history to win the U.S. Amateur and the NCAA Championship in the same year, after Jack Nicklaus, Tiger Woods, Phil Mickelson, and Ryan Moore. DeChambeau achieved his place while employing a Stroke Pattern that was based on a strict one-plane swing and a set of irons that were all the same length. With help from coach Mike Schy, the golf world was put on alert about this new age eccentric using the old book. DeChambeau continued making waves when he won two FedEx Cup playoff events in 2018 and made his first U.S. Ryder Cup team.

The objective principles of Kelley's work, using physics and geometry to break down the practical reality of the golf swing and its effects, remain true. Kelley's mission also remains intact, which states that once a player spent some time inside *The Golfing Machine* and began to understand it, then the mystery of the game begins to vanish.

"Don't turn away because the truth looks too complex," Kelley wrote. "Stay with it a while and you'll soon find it all very helpful and comfortable. After all, complexity is far more acceptable and workable than mystery is."

The 1955 U.S. Open at the Olympic Club in San Francisco might have been the strangest major golf tournament ever played. Not too surprisingly, Michael Murphy was there to see it.

Murphy had gone through a spiritual awakening when studying as an undergraduate at Stanford, leaving his original plan of following his grandfather into medicine and instead going into a directed-reading course in philosophy. He returned home with a degree in 1953, and he was soon drafted into the army for the Korean War. He spent two years stationed in Puerto Rico, mostly playing baseball and fighting what he called "the Great War of Mosquitoes, and I was like Sergeant York." After he was discharged in 1955, he returned to Northern California, thinking about going back to school for his doctorate en route to becoming a philosophy professor. While waiting for the spring semester, Murphy attended the U.S. Open at nearby Olympic, the first major played at their Lake Course.

Built on the sloping hills between Lake Merced and the Pacific Ocean, the course was lined with fifty-foot pines, cedar, and eucalyptus trees, and the rough was a dense, wide-blade rye that had been imported from Italy. With its Spanish-style clubhouse sitting atop the hill, and with cool, dense fogs often rolling in off the ocean during these summer months, it created a wistful environment unlike a normal golf tournament, especially a stuffy U.S. Open.

The main story line going in was the same as it was at every major from about 1946 through 1960—Ben Hogan. The diminutive Texan had become a mythical figure, having fought through a poor childhood when he witnessed his father commit suicide, a dastardly hook that undercut his early playing years, and a car accident in 1949 that nearly killed him and had doctors saying he would never again walk. Yet Hogan continued with his legendary work ethic, and as he continued piling up the best résumé of any golfer of his generation with almost unbelievably dramatic victories, his stature grew to enormous proportions that would never wane. By 1955, Hogan had spoken about quitting the game, the toll of preparing for and playing a four-round tournament (especially the demanding majors) just too much for his fragile body. But he did want that fifth U.S. Open title, which would have been a record.

The challengers were common: Sam Snead, the best player never to win a U.S. Open; Byron Nelson, who grew up caddying at the same downtrodden club in Fort Worth as Hogan; and Cary Middlecoff, the gentle dentist who had just won the Masters earlier that year by beating the second-place Hogan by seven shots. A player few had ever heard of was a pro out of a municipal golf course in Iowa, named Jack Fleck. A devout Presbyterian and tireless worker, Fleck had just joined the PGA Tour in January, and Hogan had been his hero. Fleck didn't have any endorsement deals, so before that year's tournament at Hogan's home course of Colonial Country Club, just outside Fort Worth, Fleck drove out to the plant where the Hogan golf clubs were made and asked for a set of irons. Hogan gave them as a gift, and they were the only two players in the field at Olympic playing those clubs.

As the tournament began, three separate events happened to Jack Fleck that he considered to be divinely inspired. In a practice round, an elderly gentleman who had been following him for much of his forty-four-hole-per-day regime pulled him aside and asked if he ever prayed to win golf tournaments. Fleck said no, and the man told him to just pray for the strength to compete, so Fleck did. He never found out who that man was. In Friday's second round, Fleck, an inconsistent putter, got what he called "a feeling" in his hands that helped him make some big putts en route to a splendid 69. Then on Saturday morning, tied with Hogan and one shot off the lead before the thirty-six-hole final day, Fleck got the spookiest sign of all. As he was shaving, he heard a voice come out of the mirror: "You're going to win the Open." He dismissed it, and it came again, louder: "You're going to win the Open!"

What happened on the golf course over the next two days was almost as inexplicable. Hogan shot a final-round 70, getting in with a total of 287, which seemed insurmountable. "If I win," he told the press while sipping a Scotch and soda, "I'll never work at this again. It's just too tough getting ready for a tournament. This one doggone near killed me." Fleck made a bogey at fourteen and dropped two back, but birdied the fifteenth to get back to one. He then hit his approach on eighteen to eight feet and dropped the birdie putt, forcing an eighteen-hole playoff the next day.

A huge cheer went up as Fleck sank the tying putt, and Hogan quietly cursed in the locker room. He composed himself and went out to meet his wife, Valerie, who had diligently been sitting on a couch on the second floor of the clubhouse all afternoon, an attendant sporadically coming in to update her on what was happening on the golf course. Through some family connection, Murphy had been inside and sitting on the same couch with Valerie for much of the afternoon, watching as she reacted to news. Now, Murphy saw Hogan limp in to collect his wife—cleaned up, hair combed, sports jacket on, and smiling. Murphy thought he saw an aura around Hogan the whole week and would later write about it. At that moment, the aura was mesmerizing.

Just before the playoff the next morning, Fleck approached Hogan in the locker room. He told Hogan he had been driving out of El Paso in February of 1949 when he saw two motorcycled police-men speeding in the other direction. He then read in the next day's newspaper that Hogan had been in an accident, and Fleck said that he had prayed for him.

"So good luck," Fleck said, "and whatever happens out there today, Ben, you'll know what I mean."

Hogan just stared back and finally replied, "Thanks, Jack. Good luck to you, too."

Fleck was a nervous wreck for the first couple holes, and after skulling a bunker shot on the par-3 third, he walked by Hogan and apologized for his poor play. "Jack, take your time," Hogan told him. "Don't worry about a thing." Hogan later said that was a strategic mistake because Fleck never missed another important shot the rest of the round.

After ten holes, Fleck was three shots up. By the eighteenth tee, the lead was one. Hogan pulled driver on the uphill, amphitheater par 4, when somehow his right foot—even with a custom-made extra thirteenth spike—slipped. He hit it on the neck, and it was a hook. His ball was buried in the left rough. For one more piece of serendipity, at dinner the night before, Fleck had been approached by a cartoonist named Cal Bailey and was presented with a drawing showing Fleck holding a sickle and Hogan in the deep rough off eighteen. Now it was Hogan with the sickle in his hand, swiping at his ball once with a sharp-edged sand wedge, only able to move it a few inches. Another swipe, and it went a couple yards. One more and he was back in the fairway, and after his fifth shot he was on the back of the green. The thick breeze blew in off the ocean and seemed to make the trees moan, the only sound heard above the murmur of the disbelieving crowd. Fleck hit his second shot on the green, and after Hogan made the 40-footer for double bogey (because of course he did), Fleck two-putted to win his only major title.

Hogan walked over and they shook hands. As the cameras surrounded them, Hogan took off his famous white cap and fanned Fleck's red-hot Bulls Eye putter.

"I'm through with competitive golf," Hogan said after, stunning the reporters even if they had been expecting this for a long time. He said he might try his hand at another Open, but "from now on, I'm a weekend golfer."

Witnessing that tournament had a deep impact on how Michael Murphy viewed the game. At the time, he was "on fire" with spiritual awakening, and Hogan helped him see that change translate to golf. Murphy had played on the team at Salinas High School and was a 4-handicap by the time he was sixteen. He said he used to play Pebble Beach for $5, and he once lost to Ken Venturi in the final of the Northern California Junior Amateur. But golf was a hobby, while this new exploration into the spiritual world was Murphy's main focus.

Soon after that Open, Murphy enrolled in graduate classes in philosophy at Stanford, but found them distasteful. They were teaching left-brain concepts such as analytic philosophy and logical positivism. So Murphy picked up and left in the spring of 1956, heading to Pondicherry, India, to study and meditate at the Sri Aurobindo Ashram. On the way, he stopped in Scotland and played one last round of golf at St. Andrews Old Course, the place where many say the game started and certainly the place where it flourished in the 1800s. He remembered that his rental clubs were "awful," and his lone playing partner was totally forgettable. Murphy did remember he shot 86, and he planned for that to be his goodbye to the game.

"Like St. Augustine saying he was going to give up sex for God, but, 'Please, God, don't make me give it up right away,'" Murphy said. "That was like me and golf. I was on a quest, and I didn't know if I was ever coming back. I might find enlightenment."

That is the truth of Murphy's life, a truth that he blurred when he sat down in 1970 to begin writing *Golf in the Kingdom*. The text is in a style that might be considered gonzo fiction, following in the footsteps of a writer that the Murphy family knew from back in the mid-1950s, Hunter S. Thompson.

"It was in that spirit of jocularity, but serious at the same time," Murphy said. "Some people take it as a satire, but it's real. It's real because, as I learned over the years, all of it happened to somebody."

The main character is named Michael Murphy, and while on his way to study in India, he stops at a Scottish golf course not-so-subtly called Burningbush. In Part One of the book, the character Murphy goes to the first tee and meets a sage Scottish pro named Shivas Irons, who is about to give a playing lesson to a student, Mr. Balie MacIver. The two are deeply serious about the game and are dismissive of Murphy's competitively American attitude about the need for success. Early in the round, Shivas Irons begins to rib Murphy for his anger and frustration with his poor shots.

"Ye try too hard and ye think too much," Shivas tells him. "Why don't ye go wi' yer pretty swing? Let the nothingness into yer shots."

The advice soothes Murphy, and before he knows it, he is playing better. He begins to smell the heather, and the sea. Then he hears Shivas and MacIver begin to talk about what they call "true gravity"— their metaphysical theory based on energy fields that surround the person, the club, the ball, and all their surroundings. Murphy tries to see these fields, and for a while he cannot. But as his game improves, he starts seeing a violet orb around his ball flying in the air, and then a yellow light around a seagull swooping in from the ocean. He thinks it's his retina playing tricks on him and tells Shivas.

"Noo, Michael," Shivas says. "When ye think tha' maybe it's yer retina, ye'r just one step awa' from really seein' things."

Over the next twenty-four hours Shivas and Murphy go through a fantastical series of events in which they tread through different states of being, from small and rigid hourglass shapes to enormously tall and flexible giants. They see auras that expand and contract, and lines of psychic energy that enter and exit the body. After the round, they go for a dinner party of Greek allusion at the house of a married couple, Shivas's friends the McNaughtons. After much whisky and talk about golf and love—all types of love—Murphy and Shivas leave to go back out to the course in the middle of the night.

In the dark, Shivas takes Murphy to a cave underneath the valley on the thirteenth hole, where Shivas's teacher lives. Seamus MacDuff is a supernatural academic hermit of half-African descent with a thick white beard, who is studying the theories of true gravity. MacDuff's cave is empty, but Shivas finds the hermit's ancient shillelagh and an old featherie ball. After much debate, he goes up to the long par-3 hole and, over the chasm named Lucifer's Rug, makes a hole in one in the pitch black.

Before Murphy leaves, he goes to Shivas's apartment, which is filled with philosophy books by the likes of Ramakrishna and Plotinus, along with Hogan's first instructional work, *Power Golf*. Shivas has also kept extensive journals throughout the years, where he begins to explain his past. He experienced a spiritual revelation when he was nineteen years old—the same age that the real Murphy had his revelation at Stanford. "He had resolved to become a priest of the church," Murphy wrote, "and start a revolution of the mind."

Shivas had been a great young golfer at Burningbush, and the wealthy older members wanted to send him to America to play on the Tour to show that Scotland could still produce champions. But, as Murphy wrote, Shivas had begun seeing the world "as a vast illusion, 'a play of shadows against the immensity' he had seen. During those months, he saw the truth in the Indian view of the world as *maya*, pure illusion; he knew what the old mystic had meant. Why return to this brawling pit when other worlds held such promise of peace and delight?"

Shivas had eventually found MacDuff staring at a tombstone in an ancient cemetery—an allusion to the grave of Old Tom Morris and Young Tom Morris at St. Andrews—and when Shivas approached, MacDuff spoke to him without turning from the graves: "You and I have much to do, for these are our final days." Shivas dropped out of college, and their collaboration began. MacDuff produced theories of true gravity, and Shivas tested them on the golf course, with MacDuff following on a white Shetland pony.

Part Two of the book consists of partial excerpts and explanations from the journals taken from Shivas's apartment while he was in a trance so deep he once almost physically disappeared. With the narrative now behind you, this is the heart of the philosophy of the book, spelled out in small subsections with titles such as "Universal Transparency and a Solid Place to Swing From," "Relativity and the Fertile Void," and "The Higher Self."

It begins anthropologically, with Shivas writing in his journals that "a round of golf partakes of the journey, and the journey is one of the central myths and signs of Western Man. It is built into his thoughts and dreams, into his genetic code. . . . But other men have not been so concerned to get somewhere else—take the Hindus with their endless cycles of time or the Chinese Tao. Getting somewhere else is not necessarily central to the human condition."

Shivas believed something inside all humans understands "we are the target as well as the arrow." For that reason, he thought, many are drawn to golf, because, as Murphy explained, "if it is a journey, it is also a *round*: it always leads back to the place you started from. . . . 'By playing golf,' [Shivas] said, 'you reenact that secret of the journey. You may even get to enjoy it.'"

Some of the subsections are vignettes and some had graphs and predictions for the next steps in human evolution. But just as with a "flat left wrist" or "sustain the lag" from *The Golfing Machine*, some flashpoints resonate more than others. Those who love *Golf in the Kingdom* will often offer a pre- or postround toast while quoting from the small subsection titled "Against our ever getting better."

"'Tae enjoy yersel', tha's the thing,' [Shivas] said, 'and beware the quicksand o' perfection.' He then raised his glass of whisky up and shouted, 'I say fuck oor e'er gettin' bitter!'"

Some sections are on how golfers are drawn to the whiteness of the ball, to the mystery of the hole, and to just watching a projectile fly through the air. Other sections are on why scorekeeping is important, as is following what can be a convoluted and complicated system of rules. "The best players come to love golf so much they hate to see it violated in any way," Murphy wrote.

These proverbs drew people to the book in far greater numbers than to *The Golfing Machine*. The fact there was some publicity helped, as well, starting with a glowing review from the novelist John Updike in the *New Yorker* in 1972.

"Golf is of games the most mystical, the least earthbound, the one wherein the walls between us and the supernatural are rubbed thinnest," Updike wrote. "Like a religion, a game seeks to codify and lighten life. Golf . . . inspires as much verbiage as astrology. . . . A golf classic if any exists in our day."

Updike had gotten the book from colleague and famed golf journalist Herbert Warren Wind, who would also sing its praises.

"A high-flying, metaphysical exercise that might be called a contemporary, Merion-blue-grass version of *Sartor Resartus*," Wind wrote. "Murphy writes with a fine knowledge of golf and with style and humor, and *Golf in the Kingdom* may very well be the best 'serious' golf book since Arnold Haultain's *The Mystery of Golf* in 1908."

The book has since sold well over a million copies and has been translated into nineteen languages. The film rights were first bought by Clint Eastwood in the early 1990s, but he could never get a proper script made that was commensurate with the power of the book. Eventually a small-budget movie was released in 2010, which, as expected, was a poor representation of the original work and was widely panned by critics.

There is even a section in Part Two titled "Hogan and Fleck in the 1955 U.S. Open." The real Michael Murphy was there, but he wrote in this section that "Shivas found himself sitting on a couch upstairs with Hogan's wife—synchronicity having led him there." Murphy wrote that Fleck had discovered Hogan's "secret," but he wouldn't share it, although Shivas could see what it was: "It was Hogan's own presence, communicated directly to Fleck. His 'inner body and command of true gravity' had emanated directly to the younger professional to such an extent that Jack Fleck won the 1955 Open title."

This blurring of the line between reality and fiction—especially in light of knowing that the real Murphy was there—helps connect people so deeply to the message. The book may have been written "in a spirit of jocularity," but it is divulging a philosophy about the game (and about the world) that is as serious and concrete as anything in *The Golfing Machine*.

"It is good to remember that the mental states described in this book are familiar to golfers the world over," Murphy wrote. "And I thought it might reassure you to know that Ben Hogan has been involved in such matters all along."

# TWO

# GOLF AND RELIGION

Michael Murphy worked on an egg-white omelet with spinach and mushrooms, two buttered slices of sourdough toast, some tea and some tomato juice, and remembered the exact moment when he became convinced Tiger Woods had a shamanistic gift.

The moment had come months before, and Murphy recounted it with guile as he sat outside on the deck of a small breakfast joint north of San Francisco. He sat tall on a cushioned wrought-iron chair with his legs crossed, his hands in his lap, and his quarter-zip sweater pulled up just below his chin. Even at eighty-two years old, he was six feet two and powerfully built. His ruby Irish face shone and his loud cackle of a laugh cut through the morning's din. To sit across from Murphy was to be engaged by a man whose intellect and intuition were utterly convincing in their breadth. He leaned forward to say something, and you leaned forward to catch it. He told a joke, and you laughed as if he let you in on a secret.

He began telling the story of recently sitting on the couch in his living room not too far from here, watching the final round of the Honda Classic from PGA National in Palm Beach Gardens, Florida. Tiger Woods was less than a year removed from his self-induced hiatus following a very public and salacious divorce, and he had yet to win a full-field tournament. Public curiosity about him, about the state of his game, and about the state of his mental fortitude was immensely high. He came into the final round a full nine shots behind leader Rory McIlroy, but by the final hole Woods had played himself back into contention. He stood in the first cut of rough just left of the fairway on the 556-yard, par-5 eighteenth hole already at 6 under for the day, just a couple back of McIlroy. Woods had 218 yards to the hole, the ball slightly below his feet, the wind into him and bit off the left. The hole was cut on the extreme right portion of the green, on a finger of land that sat out over a canvas of dark and murky water, a pile of rocks, and a deep sand trap.

Murphy sat up. "I just get in a trance watching him," he said. "Just stoned."

Staring at the screen, an image popped into Murphy's head. It was that of a younger Tiger, a dark and baggy maroon shirt buttoned to his neck and billowing around his thinner frame. Woods was twenty-four years old, and he was in a fairway bunker at the 2000 Bell Canadian Open at Glen Abbey. From that bunker, he hit one of the more memorable shots of his career, a 6-iron over a pond to a back-right pin position, locking up the final leg of the Triple Crown—the U.S., British, and Canadian Opens—as well as his ninth victory on the PGA Tour that season. Some think of that moment, above all the major-championship victories and impressive runaway wins, as being the one that defined his peak, that defined his powers at their height.

Now at PGA National, Tiger pulled back a 5-iron and hammered one just right of the hole, a marvelous shot by any account, and he rolled in the 10-foot eagle putt for a final-round 62. It wasn't enough to catch McIlroy, but when Tiger went on TV after signing his score-card, he said something that made Murphy shiver.

"For some reason, I kept thinking, 'This is very similar to what I had at Glen Abbey,'" Woods said. "But at Glen Abbey, I wasn't firing at the flag, either. I was firing at Grant Waite's ball and was just going to move it to the right, and this was the same thing—aim at the tunnel, I'm going to lean the shaft to try and take some loft off of it, and it's going to start a little further right, but just rip it.

"And I absolutely just killed this 5-iron."

Murphy remembered hearing that and then letting out a laugh that startled his wife in the other room. It was the deepest transcendental moment he had while watching Woods play golf on television. Murphy was not the only one watching to recall what happened in Canada over a decade before, but not a lot of coincidence goes on in Murphy's head, especially not concerning someone with the presumed psychic powers of Woods. Murphy has spent the vast majority of his life striving to put together mystical pieces, so he often sees and feels connections that might seem fantastical to others. He had never met Tiger Woods, but Murphy felt he knew him well.

"Somehow I've grown to care about him and root for him, and it verges on pathology," Murphy said. "In a way, he's like Frodo Baggins, on this insane thing, trying to win eighteen majors. And it's in real time. Not only is he great, but it's this quest. He's pulled the arrow way back, and that tension can sometimes be too much. But for all those reasons, I've gotten invested in him and I'm tuned in to him in some weird ways."

This was not just a whimsical experience for Murphy, but more a glimpse at the way he interprets the world on a deep level. Before and after writing *Golf in the Kingdom*, Murphy had dedicated his life to exploring a phrase borrowed from Aldous Huxley: *human potentiality*. It was described in *The Doors of Perception*, published in 1954 and fueled by Huxley's hallucinatory experience with Indian mescaline. Huxley theorized that the survival instinct in human evolution resulted in a restricted perception of the world, and this, along with advancing languages helping to shape an imperfect description of our surroundings, obscured from us some basic facts of reality. But there were avenues to regain some of that perception, and that possible rebirth was the "human potentiality."

Huxley thought we were all connected to what he called the Mind at Large. Being able to expand our consciousness gets us closer to a connection with that Mind and brings us closer to the true nature of experience—which is far from our everyday experience.

"Every individual is at once the beneficiary and the victim of the linguistic tradition into which he has been born—the beneficiary inasmuch as language gives access to the accumulated records of other people's experience, the victim in so far as it confirms him in the belief that reduced awareness is the only awareness and as it bedevils his sense of reality, so that he is all too apt to take his concepts for data, his words for actual things," Huxley wrote. "That which, in the language of religion, is called 'this world' is the universe of reduced awareness, expressed, and, as it were, petrified by language."

These are not just far-flung ideas born out of hallucination, but the basis of most religion—this world is not all that we see; our language is not satisfactory to describe the divine. Yet the rhetoric Huxley uses is in line with the Eastern philosophies of Buddhism and Hinduism and their offshoots, based on the interconnectedness of all eternal souls, and advancing the physical state of being for the betterment of the individual and the group that eventually leads to full consciousness. Western monotheisms are more focused on the singular God who wields power over his creation from on high—albeit often a loving power and a God that is striving to reunite with the subjects. The only time of full consciousness, then, is reached with a reuniting with

God at death. As Christianity rose to world prominence over the centuries, the Eastern philosophies were somewhat marginalized. They are still explored by Western intellectuals, mostly in a sense of academic philosophy. But Murphy was never for the academic, instead trying to push forward and advance what he saw as a slowdown—if not an entire stoppage—concerning the evolution of our consciousness. He thought of this not only as an individual journey, but as a group journey with grave importance for our future as a species.

So after leaving the breakfast spot, he was back at his house nearby, looking around his office for a few books that could shed light on the topic. The dark-wood house sat under a high and dense canopy of trees, with a meandering stream running underneath his front deck. He and his wife, Dulce, had to walk down the one-lane, semipaved road about half a mile to see direct sunlight. On the opposite side of the street, a steep hill of wet dirt rose some fifty feet, and blond, middle-aged women in spandex walked their golden retrievers up and down the road. The floor plan was open and neat, with high ceilings and skylights. The first floor office was a marvel, with a large desk that looked as if pieces of an ancient redwood were shaped and fitted without a nail or screw to be seen. A sleek new computer sat on it, along with a small lamp, and some neatly staked papers. A window looked out onto the mossy trees and damp underbrush.

The walls were completely made up of smooth-wood bookshelves. The room had an offshoot that turned into a three-walled enclave the size of a small bedroom and felt more like a small library. The books were impeccably arranged, and Murphy had some sort of mental Dewey decimal system that organized them by category. He walked slowly over to a far wall, squinted under his bushy gray eyebrows, and reached easily to remove one. The tiny paperback was titled *Breakfast at the Victory: The Mysticism of Ordinary Experience*. Murphy explained how the author, James P. Carse, was a renowned professor of the history and literature of religion at New York University, and this was his most popular work, based on his philosophical analysis of his time at an East Village luncheonette.

"If God exists beyond all the heavens, then God must be hidden in what is closest and most familiar to us," Carse wrote, before quoting an ancient Chinese philosopher, Chuang-tsu. "'When there is no more separation between *this* and *that*, it is called the still-point of Tao. At the still-point in the center of the circle one can see the infinite in all things.'"

Looking for another book, Murphy walked out into the main room and found it sitting on the kitchen counter. On the cover was a drawing of a magnifying glass, looking at a hand holding a shard of mirror, which in it showed an open eye. Murphy had been talking about this book all day, especially when the idea was brought up that golf is in the midst of this transformation from ethereal pastime to scientific laboratory.

The book is called *The Master and His Emissary: The Divided Brain and the Making of the Western World*, and its author is a British psychiatrist and neuroscientist named Iain McGilchrist. In comparing the creative right side of the brain and the analytical left side, Murphy described the book by saying, "The right brain is the master, the left brain emissary. The right brain in virtually everybody is bigger, it's longer, and it's covered much more with a white, myelin sheath with all the neurons that means it can compute with an order of magnitude way beyond the left brain. So the big question is, Why the hell did evolution do it this way?"

McGilchrist explains it all through dense neurological and anthropological research. He first lays out the history and the evolution of the brain, tracking the changes in ancient skull measurements up to more sophisticated measurements of the twentieth century. He then follows with the corresponding history and evolution of mankind. As the left hemisphere increased in size and power, the culture of mankind became more antagonistic, less creative, and has embraced an existential absolutism.

"The Master [the right side] needs to trust, to believe in, his emissary [the left], knowing all the while that that trust may be abused," McGilchrist writes in the conclusion. "The emissary knows, but knows wrongly, that he is invulnerable. If the relationship holds, they are invincible; but if it is abused, it is not just the Master that suffers, but both of them, since the emissary owes his existence to the Master."

McGilchrist used modern neuroscience to support a premise that is unsettling and controversial, making the leap to say that the consequences of our evolution have not only filtered out large portions of the nature of reality to make it possible for us to survive and procreate, but that evolution has changed who we are to slowly turn us against one another and away from the path that carried us to this place of semiconsciousness and global dominance over other species. This was the point Murphy was trying to make, that exploration of the creative and mystical and immediately unexplainable is how we have advanced to this point of consciousness. To continue that exploration is how we advance further. Thinking there is that room to improve the nature of our spiritual selves—both individually and as a species—is distinctly Eastern. It was reiterated by Huxley and Carse and countless others, and it's thought of not just as a philosophical abstract. It is an overarching idea that has resonated in people for millennia—the mysterious optimism that we have the ability to change our souls.

"The man who comes back through the Door in the Wall will never be quite the same as the man who went out," Huxley wrote about his experience. "He will be wiser but less cocksure, happier but less self-satisfied, humbler in acknowledging his ignorance yet better equipped to understand the relationship of words to things, of systematic reasoning to the unfathomable Mystery which it tries, forever vainly, to comprehend."

Ben Doyle shook as he said it, sitting outside in the cool San Francisco shade, an eighty-year-old man born on Bastille Day with not a lot of time left. It was the fall, and when Doyle was asked about golf, he began speaking about religion. For him, the two were intrinsically intertwined.

"Truth is truth, love is love," he said. "God is love. God is truth. God is soul. God is mind."

These are tenets of Christian Science, and behind Doyle was a four-story, Spanish-style building, fronted by a circular driveway and a fountain colored by light-pink tiles. When Doyle had gotten older and transitioned away from teaching golf every day, he moved here, to the Arden Wood Nursing Home, up in the hills of Wawona Street, started in 1929 under the Christian Science principles set forth by the religion's founder, Mary Baker Eddy. As Eddy told the story, things began in 1866 when she slipped on some ice and suffered a serious back injury. Through an in-depth reading of one of Jesus's healing stories in the Bible, she was ridded of her ailment.

But Eddy believed she wasn't cured because that would imply she was hurt. After deep inspection of the Bible, Eddy believed she had found a set of specific instructions laid forth by Jesus explaining how he was able to perform his healing miracles. It was based in the idea that "God is love," and that the true world he created is perfect. Therefore, all the maladies that define everyday life are products of human sin. The physical reality we live in is not an outside environment, but an illusion, a manifestation of the human internal. With pure thoughts and love, all those maladies—physical, emotional, spiritual—can disappear. With directed prayer that she outlined, all sickness can be vanquished. In later years, this obviously created quite a problem with the medical community, and after the children of some Christian Science practitioners had died, the parents were successfully prosecuted for neglect and manslaughter. Yet the church itself doesn't take the word *science* lightly.

"Christian Science is also a science because God is understood to be unchanging Love—the infinite Principle that is constant, universal, inclusive, eternal, the only true power and source of all good," reads their literature. "It explains the spiritual laws of Love that enabled Jesus to heal sickness and sin. This divine Science also answers our fundamental questions about evil, reality, and eternal life. And as the word *science* implies, it is reliable, consistent, and provable, bringing healing to individuals and humanity through a deeper understanding of God."

The institution of the religion is based in Eddy's writing, starting with her main work, *Science and Health with Key to the Scriptures*, published in 1875. The only other bedrock text is the Bible, Eddy writing that "as adherents of Truth, we take the inspired Word of the Bible as our sufficient guide to eternal life." By 1879, she and twenty-six followers were granted a charter to start the Church of Christ, Scientist. It quickly became the fastest-growing religion in the United States, peaking with nearly 270,000 members by 1936. But as medicine and doctors improved, the membership dwindled, and it was estimated that fewer than a thousand members were left in the United States by 2009.

One of those was Doyle, who was first exposed to *Science and Health* as a child growing up in Coquitlam, British Columbia, about a half hour inland from Vancouver. Doyle's mother, Catherine, ran the food concession at Vancouver Golf Club, while his father was a salesman for a log-trucking company. One day, Doyle went up to the attic of his comfortable middle-class home and unearthed an old Bible as well as *Science and Health*. He started flipping pages and liked what he was reading. He came downstairs with Eddy's book and asked his mother what it was.

"They're medical books," she said. Doyle had already decided he didn't want to be a doctor, so that was that.

Doyle already had the golf bug embedded deep, caddying at the course where his mother worked while becoming obsessed with technique. He once went two years without hitting a ball, just taking practice swings. But when he did start playing seriously, he earned himself a golf scholarship to Western Washington University in Bellingham. While there, he was also the part-time assistant pro at Bellingham Country Club, where the college team practiced.

One day, he and a friend named Pete were heading to hit balls on the Bellingham range when they made a quick stop at the tennis court to have a little fun with some local girls, do a little dancing on the court in the middle of their practice. Pete was dating one of those girls, Joanne, and when Doyle met her, he, too, was smitten. Joanne participated in Christian Science readings on Tuesday nights at Western Washington, so Doyle suddenly decided that Christian Science was now his utmost interest. He asked his mother to send him those two books from the attic, and he thoughtfully went over them and began attending the meetings.

Eventually Pete and Joanne broke up, and one of Joanne's friends told Doyle to ask her to the homecoming dance. He did, the two started dating, and soon thereafter they were married.

Almost fifteen years into his marriage, Doyle was the head pro at Broadmoor Golf Club in Seattle. One afternoon on a rainy Saturday in December of 1969, a big white Cadillac pulled into the parking lot, and Homer Kelley got out of the car with three first editions of *The Golfing Machine*. The green hardcover books had dust jackets of corn yellow. He walked into the grillroom and introduced himself to Doyle, saying they had a mutual friend from the local Christian Science reading group, Bill Thomas. That got Doyle's attention, and Kelley began telling him about the book. They didn't stop talking for six hours.

"He was a problem solver," Doyle said of Kelley. "I respected him right away. I could see what he did."

Doyle spent the night reading the book cover to cover, and it all made sense to him. Like Eddy figuring out the laws that led to Jesus's healing power, Kelley had found the laws that led to successful golf.

That Sunday morning, waiting for the security guard to open the gate into Broadmoor, Doyle began reading again. It was chapter 4, "Educated Hands." It became Doyle's favorite part, seemingly so intuitive. "The player must acquire, and continue to develop, habitually skillful, disciplined, conscious manipulation of the Hands . . . as the main line of communication between Hands and Clubhead—both ways." Doyle couldn't wait to get inside to call Kelley and tell him he *already knew* that principle, but was never able to explain it so simply—so scientifically.

"Start with principle and stay with principle," Doyle said. "Law never takes a day off."

Just like Kelley walking into the grillroom on that rainy day, a simple twist of fate changed Michael Murphy's life, as well.

Murphy had been en route to following his famous grandfather into the medical field, but on the second day of classes in the spring of his sophomore year at Stanford, he walked unknowingly into a lecture hall in Cubberley Auditorium, expecting a social psychology class. Because of a last-minute room change, what he got was a seat in front of Frederic Spiegelberg, who was one of the foremost lecturers on Indian civilizations and psychology, and the comparative spiritual similarities of the world's religions.

Growing up, Murphy was the only member of his family who went to the Episcopalian church, becoming an altar boy at the age of eleven. When he got to college, "the Virgin Mary floating up to heaven, all that stuff went away," he said. Walking out of Spiegelberg's class, what resonated with Murphy was the message from the Indian seer Aurobindo. Having been educated at King's College in Cambridge during the turn of twentieth century, Aurobindo was at first a radical political figure calling for India's independence from Britain. After some bombings were wrongly connected to his organization, he was arrested. While in jail, Aurobindo said he had mystical experiences, and when he got out, he left the political sphere to go to the countryside of Pondicherry and focus on his spiritual work. There, he established the Sri Aurobindo Ashram in 1926 and began to develop his own philosophy, which he called Integral Yoga. As described in his seminal 1939 work, *The Life Divine*, the process was to advance spiritually and eventually evolve from human life into the life divine. He wrote several additional books about his study and also published some fiction and various translations of the ancient Indian philosophical texts, including the Upanishads and Bhagavad Gita. He was nominated for the Nobel Prize twice, once for literature in 1943 and once posthumously for peace in 1950.

Soon after Aurobindo died, Murphy took Spiegelberg's class. Murphy was beginning the phase of his life when he said he was "on fire" with a spiritual awakening, and he essentially took vows to dedicate his life to Aurobindo's traditions. (One of them was a vow of chastity, and Murphy said he remained a virgin until the age of thirty-two.) When he returned home that summer, he and his brother, Dennis, were playing a round of golf when Dennis turned to him and said, "You're becoming a golfing yogi!" When Michael returned to school, he quit his fraternity, Phi Gamma Delta (the same as Jack Nicklaus), and quit premed. Because he was a good enough student, he said he was allowed to enter a course structure called directed reading, in which the student creates his own courses under the supervision of a mentor. When he told his father of his plan, Murphy was financially cut loose.

"A yogi in Salinas in 1951 was someone who slept on a bed of nails," he said.

Murphy was so adamant about his new beliefs that he went and spent three years at the Sri Aurobindo Ashram—stopping on the way to play golf in Scotland—and when he returned to California, his goal was to establish his own type of modern ashram. In India, he and his fellow *sadhaks* had meditated most of the day, with the distributed literature saying there "must be a surrender to the Divine and an opening to the Divine Force so that it may work to transform one's being." He wanted to create that same environment back in America, but also wanted it to be a place that was less concerned with enforcing rules and more concerned with expanding boundaries.

"I was stoned most of the day," Murphy said, "because when you sit that much, you don't really have to have a great mystical capacity, like Hare Krishna, but I had enough. I was on fire [without] knowing how to embody it in the world."

What Murphy did know was the perfect place to start this exploration. In 1910, his paternal grandfather, Henry, the famous doctor, had purchased about four hundred acres of Pacific coastline in the yet-untamed area of Big Sur. As shockingly beautiful as any place in America, the land of Big Sur is full of hundred-foot cliffs, ragged and rocky shoreline, bustling ocean, and towering inland forests going up into the steep Santa Lucia Mountains. At the proper times of year, blue, gray, and humpback whales can all be seen migrating south to Baja and north to Alaska.

In 1948, Michael's grandfather died. His grandmother, Bunny, bestowed the caretaking of the property, including its houses and cabins, to her "minions," as Murphy described them, fellow members from the First Church of God of Prophecy. By 1960, the scene was surreal. A "muscular gay community," in Murphy's words, came down from San Francisco and up from Los Angeles to use the natural hot springs, baths originally used by the Esselen Indians. Some beatniks hung around, such as Jack Kerouac, Lawrence Ferlinghetti, Allen Ginsberg, and Alan Watts. There were also motorcycle ruffians with a penchant for marijuana and fighting. After returning from India, Michael repeatedly asked Bunny for the land, but she resisted, telling family members, "Michael will just give it to the Hindus."

"It was a three-ring circus and she didn't know," Murphy said.

"Eventually someone told her, 'We have to give it to Michael or we'll all end up in jail.'"

In 1961, when Bunny finally became aware of the chaos on her property, she hired a twenty-three-year-old from the Kentucky farmland, fresh out of the Air Force, to act as a security guard. Hunter S. Thompson came to Big Sur armed, with no compunctions about resorting to violence or firepower. The man who would go on to create his own literary form known as gonzo journalism—exemplified in his 1971 work, *Fear and Loathing in Las Vegas*, a piece of blurred fiction that the *New York Times Book Review* called "the best book on the Dope Decade"—Thompson came to the Murphys' land by way of his admiration for Michael's brother. In 1958, Dennis had published his first novel, *The Sergeant*, about a homosexual affair in the U.S. Army, which garnered terrific praise, including from family friend and mentor John Steinbeck. Thompson spent most of his time reading that book and taking notes in the margins, fine-tuning what would become his own vicious stranglehold on the English language.

That spring, Michael Murphy returned to the property in the middle of the night, parking his red Jeep before going to sleep in the Big House, the original Murphy home on the land. He awoke to Thompson pointing the hulking barrel of a six-shooter at him.

"Who the hell are ya and what the hell are ya doing here?" Thompson barked.

As the story goes, Thompson returned to the property days later with his girlfriend and two hitchhiking soldiers from Fort Ord, just north of the Monterey Peninsula. In his wild-eyed state, Thompson decided to go down to the baths, where he began his usually loud and abrasive interaction with the gay men. When things got heated, the girlfriend and the soldiers ran, while Thompson got brutally beaten and was threatened with being thrown off the cliff into the rocky waters below. The next day, bruised and distraught, harried and inconsolable, Thompson began firing his pistol through the closed window of his room.

Coming to Big Sur to write what he called in his letters "The Great Puerto Rican Novel"—which would eventually become *The Rum Diary*, not published until 1998—Thompson struggled to find paying work. After many failed pitches, he eventually sold a piece to *Rogue* magazine, which paid him $350 in July of 1961 to publish "Big Sur: The Garden of Agony." That area of the country was becoming known as a party capital, and Thompson's piece only created more intrigue. It began:

> If half the stories about Big Sur were true this place would long since have toppled into the sea, drowning enough madmen and degenerates to make a pontoon bridge of bodies all the way to Honolulu. The vibration of all the orgies would have collapsed the entire Santa Lucia mountain range, making the destruction of Sodom and Gomorrah seem like the work of a piker. The western edge of this nation simply could not support the weight of all the sex fiends and criminals reputed to be living here. The very earth itself would heave and retch in disgust—and down these long, rocky slopes would come a virtual cascade of nudists, queers, junkies, rapists, artists, fugitives, vagrants, thieves, lunatics, sadists, hermits and human chancres of every description.
>
> They would all perish, one and all—and, if justice were done a whole army of tourists and curiosity-seekers would perish with them. All the people who come here "for a few kicks" would share the fate of the doomed residents, and anyone surviving the Great Slide would be done in by Killer Whales. The casualty list would be a terrifying document. In addition to the locals it would include voyeurs of all types, hundreds of free-lance pederasts, every sort of predatory jade, and a legion of would-be orgy-masters.
>
> None of this is likely to happen, however, because almost

everything you hear about Big Sur is a rumor, legend or an outright lie. This place is a myth-maker's paradise, so vast and so varied that the imagination is tempted to run wild at the sight of it.

When word of the shooting incident and a copy of the story got back to Bunny, it was the last straw. She got in her black Cadillac with her Filipino driver and drove down from Salinas to Big Sur, where she personally fired Thompson and told him to leave the property immediately.

Soon thereafter, she gave the land to Michael, who, in turn, gave it to the Hindus.

By 1963, Murphy and his spiritual compatriot Dick Price had come together and used the land to establish the Esalen Institute. Dedicated to the idea of "exploring the human potential," the place quickly turned into a playland for drug users and the sexually ambitious. Murphy estimates that between the fall of 1967 and the end of 1970, more than two thousand articles were published about Esalen. "They branded us the touchy-feely capital of the world," Murphy said. He remembered one newspaper headline read, "Esalen Is a Dark and Dirty Place."

The de facto agreement between law enforcement and Murphy was that he and his compatriots would be left alone as long as they stayed on the land. With omnipresent pot smoke flavoring the air and LSD consumed like a food group, Esalen became widely infamous as a party palace. (Much later, it became a retreat for giants of the tech industry, who came down from Silicon Valley to grapple with the ethics of their business.)

As Esalen grew, so did the breadth of topics covered in its workshops. Surprisingly, golf was not of much interest, despite Murphy's history. But eventually Steve Cohen showed up, and a casual conversation with Murphy led to a round of golf and eventually the founding of a new workshop titled Golf in the Kingdom.

Cohen had been a public-school teacher who grew up in the shadow of Yankee Stadium in the Bronx, and he was recruited to Esalen in the mid-1970s during a traveling workshop at the Commodore Hotel in midtown Manhattan. Having taught students he described as "disturbed," Cohen wanted to learn more about Gestalt therapy, a practice started by an Esalen tenant, Fritz Perls. At the time, Gestalt was one of the newest trends in helping mental illness. The basic premise is to rebuke Freud by saying the past can be left behind, and the therapy is "an attempt to focus on a subject's conscious experience and construction of the here-and-now (instead of on the forgotten past and the unconscious dynamics that the past produces, as Freud attempted to do.)" Progress in Gestalt is made by dealing with the past as a series of events that led to the present. The baggage from those events is only carried around if one chooses not to accept them and move on.

After the workshop in New York, Cohen had a sit-down with one of the most prominent members, George Leonard, the charismatic literary icon, the West Coast editor of *Look* magazine, and the agent for Jack Kerouac and Ken Kesey. Leonard was tall and powerful and commanding, and he and Cohen stayed up all night talking.

"I grew up in the uptight culture of the fifties," Cohen said. "It was all very exciting, and I wanted more."

Author Janet Lederman, who wrote a famous Gestalt book, *Anger and the Rocking Chair*, was a frequent visitor and contributor at Esalen, so Cohen decided he would go out to Big Sur in hopes of meeting her. When he arrived at Esalen, he found the entrance plain: a small wood sign, with a regular-looking driveway off the circuitous Highway 1, which runs along the beautiful coast. He was unprepared for what lay ahead. A steep hill led down through the trees to a small gatehouse, and past that, the land continued to slope toward the sea. After walking through tall pine and cypress trees, with some small shacks scattered about the land, he came to a plateau and a clearing.

There, the mighty Pacific, in all its enormity and elegance, spread out in panorama.

"There are things that change your life," Cohen said. "That walk was a life-changing experience."

He signed up for a two-week course called Experiencing Esalen. "It was a lightweight class," said Cohen, who would later go on to teach it. "I wanted to get deeper." So he signed up for a workshop on Gestalt, taught by a couple named Beverly and Julian Silverman. The class was in the same building Perls had called home, with a circular façade and a side bedroom where Perls had slept before he died of heart failure in 1970. The main room was a large, windowed space with a vaulted ceiling, all blond wood with pillows thrown everywhere in the place of seats or tables or furniture of any kind. Cohen sat on the floor next to the fireplace. A man named Joe, who was from Nashville, spoke to the class about his abusive father, and Cohen was so moved that he began to cry.

Julian Silverman came over to him and asked, "What's that about?"

"I'm feeling touched," Cohen said. "I'm crying for Joe."

"What would happen if you cried for yourself?"

Cohen started sobbing, and he sat there for hours with Julian and Beverly while all of this pain and anguish poured out of him as if from an open dam. He went back to New York to gather his belongings and returned to California. He took his golf clubs with him, if only because they were some of the most expensive things he owned, including his cherished Kangaroo bag. He reread *Golf in the Kingdom*, and with his new focus on Gestalt, it started to make a lot more sense. Eventually he bumped into Murphy at Esalen; they got to talking about the book, golf, and Gestalt, and Murphy told him to set up a round somewhere and fill out a foursome. Intimidated, Cohen wanted to take a lesson first. He heard of a pro up in Seaside, John Allen, who approached the game differently. Not so coincidentally,

Allen worked at a driving range dubbed the Kingdom. When Cohen got there, Allen told him to grab his putter and they would head out to the practice green. If Cohen liked what he heard, he could sign up for a set of ten lessons.

Cohen agreed, and on the putting green Allen instructed Cohen to close his eyes as he did some exercises. By asking him to guess blindly about how his body moved and positioned itself, then showing him the truth—thus illustrating humans' flawed proprioception, or sense of bodily awareness—Allen quickly made Cohen more aware of what he was physically doing. Allen also asked questions about what Cohen was thinking while putting, calling his attention to mental blind spots that limited his ability to perform under pressure. After thirty minutes, Cohen wrote Allen a check for the next ten lessons. As Allen noticed the address of the Esalen Institute printed on the check, Cohen noticed a copy of *Golf in the Kingdom* sitting front and center on Allen's desk.

Allen asked if Cohen knew Michael Murphy, and Cohen explained the whole situation.

"You know," Allen said, "I'd love to meet him. And so would my teacher." It struck Cohen as odd that a teacher with this much immediate impact would have a teacher of his own.

Allen said that he would make a date as a foursome at the recently opened Poppy Hills, a sparkling Robert Trent Jones Jr. design that wove through Del Monte Forest up in Monterey, not far from Pebble Beach Golf Links. Cohen spoke to Murphy and the date was set.

That game at Poppy Hills would be Cohen's second big breakthrough in life, following his transcendent experience walking down the hill at Esalen. The day he met Fred Shoemaker became a jumping-off point for how Cohen would spend the rest of his life.

Shoemaker was physically unintimidating, small and well built, with slightly curly light-brown hair, and an inviting German-Irish face. He was soft-spoken, calm, and neat; perpetually comfortable in his own skin. With the soft speech of a psychiatrist or a nurturing friend, Shoemaker created an atmosphere of openness, of unpretentious understanding and omnipotent empathy. Cohen knew immediately that this man had to be Allen's teacher.

Unsurprisingly, Shoemaker turned out to be one hell of a golfer. He hit the ball with an effortless grace that belied the distance he got out of his five-feet-eight frame. "I had just never seen a golf ball really hit like that," Cohen said. With Allen a pro in his own right, and Murphy still able to play in the low 80s, Cohen was the novice of the group. But the four of them could not have had more fun. Allen, the bawdy old golf bum that he was, told stories about Monday qualifiers on the Tour that made Cohen almost split his side laughing. He wouldn't find out until later that they were all made up. Shoemaker brought with him an intelligence that only Murphy could match, and the laughter was just as frequent as the shots.

After the round, all four of them went back to Allen's house. They ordered pizza and drank beer and continued talking late into the night.

"I used to work with learning-disabled kids," Cohen said.

"Well, I work with learning-disabled adults," Shoemaker answered.

"You know," Murphy said, "you guys ought to teach a workshop at Esalen."

By the fall of 1988, Cohen and Shoemaker added a description to the new Esalen catalog for a workshop titled "Golf in the Kingdom . . . an exploration of the inner game." The five-day course cost $575, plus $70 in greens fees for two rounds of golf up in Monterey. The course description advertised enlightened sessions where "teaching methods gleaned from a study of the inner game will be utilized, as well as principles and methods from psychosynthesis and gestalt therapy, to explore the inner self and how one interferes with its emergence."

In practice, the workshop was based around group rap sessions, directed by Shoemaker and Cohen. The game was discussed in depth, and Shoemaker would lead people in the direction of helping themselves. Cohen would use Gestalt as a way of analyzing why golfers came to the game in a certain state, and how they could leave their baggage behind in order to excel. They would talk before they played those two rounds, and after. They would discuss what went well and what could improve, and how to do that while enjoying the game and its intricacies and nuance. And they discussed Murphy's book—floating orbs and the notebooks of Shivas Irons. Rather than working on physical adjustments, they had detailed discussions about what it felt like to be in an unconsciously focused state. The course was a communion with the principles of the book, not a physical reenactment.

Attending the first workshop was Andy Nusbaum, who ran the golf schools for *Golf Digest* magazine. The publication was a paradigm of golf's conservative past. But when Nusbaum returned, he wrote up a glowing review of his experience in the most popular golf magazine in the world, and from there, the workshop's popularity took off.

"It was using golf to learn about yourself," Cohen said. "We were teaching awareness. Know where things are, then you have some control. That's exactly what Gestalt process is. When I had that moment with my tears coming down, I became aware of my own pain. Then I can change my behavior. Then I can let it out. That's what I do when I teach and I facilitate Gestalt: I help people become aware of the baggage they're carrying around, and they don't have to. But you can't put it down if you don't know you're carrying it.

"Those moments of awareness happen," Cohen said, snapping his fingers twice. "So that's what I was teaching, and that's what Fred was teaching in golf."

After a childhood in a navy family that moved all over the

world, Shoemaker found his teaching roots at UC Santa Barbara, where he was accepted on a golf scholarship and eventually took over as the team's head coach. Wanting to know more about being a coach, he went to the Lobero Theatre in Santa Barbara to see a lecture by Tim Gallwey. Gallwey had just published a successful book called *The Inner Game of Tennis*, and soon enough he would expand his "inner game" philosophies to golf.

As Gallwey spoke, Shoemaker sat rapt. "Sometimes you don't know something is seminal to you until you look back on it," he said. "It was a talk that—I'm not sure it changed everything, but it may have."

In a short sequence of deduction, Gallwey broke down most preconceived notions about self-awareness, starting with the question "Who are you?" He went on, "You're not your mind, because your mind can change all the time. You can change it right now. You can be emotionally engaged in something and then just stop it and do something else. And you're not your body. If you had to lose a finger or a hand, you're no less you. And you're not your emotions, because if you're feeling a certain way emotionally, in three seconds if someone came running down the aisle screaming 'Fire!' you'd feel another way.

"So if you're not your mind, you're not your body, and you're not your emotions, then what are you? Thank you very much for coming."

Shoemaker was so impressed with Gallwey's ideas that he asked to see them in action. He would go to observe Gallwey giving tennis lessons and watch as his students transformed not just their games, but their outlooks on the world. The two men started to play golf together, and although Shoemaker does not think what they had was a partnership, soon thereafter Gallwey published what would become his bestselling book, *The Inner Game of Golf*.

At last, the day came when Shoemaker experienced Gallwey's transformative teaching firsthand, opening himself up to take a golf lesson. Shoemaker had known since he was a kid that he had a hitch at the top of his swing where he would regrip the club and shut the face slightly. Gallwey asked him a series of simple questions, the first being, How do you know that's true? Shoemaker replied that his friends saw it, he saw how his divots changed, and he had even watched it on film. Then Gallwey asked if Shoemaker had actually experienced it in the moment, and he had to answer honestly. He said no.

Gallwey then asked if Shoemaker would consider just paying attention and not trying to fix it. Just focus and observe what he was doing. "How do you know it's the wrong thing?" Gallwey asked. "You've never felt it."

Shoemaker estimates that he had taken nearly half a million golf swings in his life up to that point. Within twenty minutes of this lesson, for the first time he began to feel his fingers lift off the club, shift position, and reset. Since that moment, he said that problem has never again happened.

"That was a big deal, about how awareness, the simple act of being present to something over time, will develop people," Shoemaker said. "It's the only thing that is curative."

After a quick stint in the Peace Corps, Shoemaker returned to California and began teaching at local driving ranges. As his client list grew, he met John Allen, who introduced him to Steve Cohen, who introduced him to Michael Murphy. They started the workshop, and Cohen and Shoemaker also started an offshoot, a traveling band of like-minded people called the Shivas Irons Society. Shoemaker realized he needed more than a one-hour lesson to impact people, so he turned his focus toward weeklong golf schools, which he called Extraordinary Golf. He wrote a few successful books—calling them "instructional" would be to shortchange their message. He taught seminars in Florida, South Carolina, and Connecticut, along with Japan, Uruguay, Argentina, and Ireland. Fortune 100 companies approached Shoemaker for private getaways, and he began coaching the titans of industry, teaching them just as much about life as about golf.

"There is no '*is world*,' " Shoemaker said. "It's simply a question of each person interpreting things for themselves."


Ben Doyle wrapped up his Facts and Illusions mat at Arden Wood and rode up to Presidio Golf Club, in north San Francisco. This was where he felt most alive, standing behind a student and instructing. The time was long behind him when he would teach 364 days a year at Carmel Valley Golf and Country Club (later known as Quail Lodge), taking the day off only on Christmas, when he would play Pebble Beach with his children. Even when he retired from teaching year-round, students would still seek him out for private lessons, but those became few and far between as Doyle got older.

Yet the energy rose in Doyle as he shuffled toward the student on the plastic mat, telling him to chip a 6-iron.

"Shorter, slower, heavier," Doyle said. The student hit one 15 yards, then 50 yards, then 140 yards, then a full and sturdy 185. One went left, flushed. One went right, thinned. Doyle just watched, silent. Nearby were his daughter, Suzie, and her son, Bentley—which was also Doyle's full first name. Bentley had taken after his grandfather, and he was teaching at Quail Lodge after a stint at Forest Hills Country Club in St. Louis and a trip to teach at the Bhutan Youth Golf Association with Rick Lipsey. The two Doyle men, two generations apart, looked strikingly similar, and they could converse in numbers that cited chapters and subsections from *The Golfing Machine*.

When asked what first drew him to golf, the elder Doyle simply said, "Perfection." When asked why players educated in the physics of the game still have their ups and downs, he could only muse, "It's a very precise game, a very precise game. Look, I don't know what happens."

In the spring of 2012, Doyle traveled to St. Louis to watch his twin granddaughters graduate from Principia College, the only Christian Science college in the world. Joanne hated planes, so she stayed home. Doyle came back to Arden Wood alone and said hello to the receptionist in the lobby. He walked past some gold lettering hanging on the wall, spelling out a quote from Mary Baker Eddy: TO THOSE LEANING ON SUSTAINING INFINITE, TODAY IS BIG WITH BLESSINGS.

He kept moving and loaded into the elevator. He got out at the second floor, made a left, and went down the hall. He opened the door of the joined room he and Joanne shared. He walked into the bedroom. "She was just beautiful," he said, sobbing. "Laying there, just gorgeous."

Joanne Doyle died on May 13, 2012. It was Mother's Day.

On December 15, 2014, Ben joined Joanne, going quietly in his sleep at the age of eighty-two.

The message of *The Golfing Machine* did not expire with Doyle, those objective tenets of the physics of geometry that give a road map to a complicated motion. Nor did the analogous idea of a perfect world end with the decline of Christian Science. People were drawn to Eddy's church by the comforting knowledge that the world God created is perfect, and the closer you get to the divine, the closer you get to perfect. Golf being such an intricate and complicated game, it is tempting to gravitate toward concrete answers. It is a lot easier than, as Shivas Irons said, "Let the nothing into yer shots."

Murphy's shaman, Tiger Woods, spent a good part of his life chasing perfection. As his career progressed, Woods became fixated on finding evidence to prove that he was playing at his very best— evidence beyond the sweeping victories and shattered records in his wake. By 2012, he was obsessed with the numbers being spit out by a precise ball-flight monitor known as TrackMan.

"Understanding those numbers," Woods said, "is relevant because it's pure numbers. There's no getting around it. They're universal. They're law."

As he always did, Woods kept pushing for more, kept pushing to get better. He wanted to fully understand what was happening. When his father, Earl, died in 2006, it seems the rudder of Tiger's life disappeared. By 2009, he was spiraling out of control, and late on that Thanksgiving night, he drove his Escalade into a fire hydrant in his gated community in Florida. His plethora of extramarital affairs had finally come to light, and his Swedish model wife, Elin, chased him out of the house and smashed the back window of the car with a golf club.

The sensational scene created a whirlwind of gossip, and the *New York Post* featured the scandal on the front page with eye-popping tabloid headlines for twenty consecutive days, one day longer than the September 11 attacks had held the cover.

But perhaps the most telling detail from the incident was a book found on the floor of Woods's car. There among the shattered glass and deflated airbags was a small paperback titled *Get a Grip on Physics*. Written by British scientist John Gribbin, the book is a quirky introduction to the basics of physics, starting with Newton and Galileo, to Schrödinger and Einstein, eventually breaking down quantum mechanics, string theory, and the ever-evolving understanding of the fabric of reality. After all the facts of the accident came to light, sales of the book immediately skyrocketed.

Why did Tiger Woods, a man with such enviable talent, both physical and mental, a man who for so long had the world in his palm, have this book in his car in the first place?

# THREE

# TIGER WOODS AND DUALISM

Modernity is essentially a dualist state of being, and Tiger Woods is nothing if not golf's modern man.

As science backed by empirical data has variously assaulted or confirmed our sense of human history and our intrinsic understanding of nature, it is not necessarily a moral dualism that has emerged, but rather an ontological dualism—the philosophies concerning the nature of being have become polarized. Although some philosophies and religions still have a straightforward moral framework built on clear ideas of good and bad, creator and destroyer, the modern conflict is not so clear-cut. The artist faces off with the scientist, the creative dreamer vies with the grounded pragmatist. It seems foolhardy to shun all technology and its advances, disregarding new discoveries as blasphemous. Yet it seems equally shortsighted to believe that human existence can be fully understood by compiling enough data.

Bridging the gap between the subjective and objective creates a balance. In golf, this neutral ground is a place from which excellence can be born. The combination of feel and technicality has always been present in the game. No great player has ever thought of him- or herself as "technical," but everyone has had an individual understanding of his or her technique. Everyone longs to be the gifted artist, but when it comes to performance, most people need to wrap their hands around something tangible. Homer Kelley wrote he wanted "Feel to come from Mechanics," which is a lot easier for players to grasp than the underlying meaning of Shivas Irons's theory of true gravity. It's also a lot easier to practice the mechanical, and a lot easier to see tangible improvement.

Yet in Woods's deepest moment of peril, he publicly conceded that he had become too concerned with the material world and had strayed from a spiritual balance that had helped him achieve such greatness. On February 19, 2010, just about three months after his car accident and his affairs made headlines around the world, he stood in front of a lectern at the PGA Tour headquarters in Ponte Vedra Beach, Florida, and delivered a poignant televised speech. It was as close to a stripped-down version of himself as he had revealed in a long time, devoid of sponsors and excessive doublespeak. He seemed sincere in his apology to his family, friends, and business associates. He took responsibility for the collapse of his marriage and the demise of his reputation, saying, "I recognize that I have brought this on myself." At the end of the fourteen minutes he said, "I ask you to find room in your hearts to one day believe in me again."

When discussing how he intended to improve, Woods focused on something that few people saw coming.

"Part of following this path for me is Buddhism, which my mother taught me at a young age," he said. "People probably don't realize it, but I was raised a Buddhist, and I actively practiced my faith from childhood until I drifted away from it in recent years. Buddhism teaches that a craving for things outside ourselves causes an unhappy and pointless search for security. It teaches me to stop following every impulse and to learn restraint. Obviously, I lost track of what I was taught."

Woods was wearing two small rope bracelets on his left wrist, and he later explained why: "It's Buddhist. It's for protection and strength. And I certainly need that."


From the beginning, Woods harbored an internal struggle, and it started with his parents. His father was an African American with mixed ancestry, a military man from Manhattan, Kansas. Earl was a Green Beret in the Army Special Forces and had three children from a previous marriage. While stationed in Thailand in 1966 during his second tour in the Vietnam War, he met a woman named Kultida, twelve years his junior. Tida, as she was known, had Thai, Dutch, and Chinese ancestry. She and Earl were married in 1969, and she gave birth to a son, Eldrick Tont Woods, on December 30, 1975. Earl reportedly nicknamed him Tiger after a fighter pilot in the Vietnamese army, Vuong Dang "Tiger" Phong.

Earl earned a reputation as hard-nosed, testing young Tiger with mental obstacles in hopes of making him stronger. Earl would throw clubs on the ground as Woods was swinging. (Ever seen Tiger stop midswing and wonder how he does it?) Earl would turn good lies into bad lies, toss balls into bunkers, move par putts from three feet to ten feet. When Tiger was three years old, he hit a ball behind a tree and Earl made him describe all of his options, beginning to explain the concept of course management. Earl would drop little philosophical sayings such as "You get out of it what you put into it." Woods said these phrases would "fester" in him and eventually manifest in his life on and off the golf course.

Tida was diminutive in stature but a disciplinarian at home. Neither she nor Earl would scold Woods for playing poorly in tournaments, but they would scold him for not acting properly, both on and off the course. Woods said there was "zero negotiation" with his mom, and that was part of the reason that Buddhism became his religion.

"As we said in our family, my mom was the hand, and my dad was the voice," Woods wrote in his 2017 book, *The 1997 Masters: My Story*.

The family lived on Teakwood Street in Cypress, California, and Tiger grew up as an only child. Earl and Tiger would head to the nearby Navy Golf Course to play and practice, then spend the afternoon sitting in the grillroom talking with the old military men while fighter jets landed on a strip adjacent to the eighteenth fairway. Earl would order a rum and Diet Coke, Tiger would get a Coke with cherries, and the men would reminisce about the military while Tiger sat and listened, occasionally asking questions. It was the beginning of Woods's lifelong obsession with the military.

When he was two years old, Woods was famously brought on *The Mike Douglas Show* to hit balls into a net, presented as a golfing prodigy. By the time he was fifteen and attending Western High School in Anaheim, Woods had won three consecutive U.S. Junior Championships. By the time he was sixteen, he played in his first PGA Tour event, missing the cut at the L.A. Open. In the summer of 1994, Woods won the first of his three consecutive U.S. Amateur Championships, fist-pumping his way around TPC Sawgrass in a straw hat and above-the-knee shorts. Woods had been six down just thirteen holes into the championship match, but won the last three holes to beat Trip Kuehne and complete what's believed to be the biggest comeback in the history of the Amateur final, going back to 1895. A month later, Woods enrolled in Stanford, where teammate Notah Begay III nicknamed him Urkel for his skinny frame and overall nerdiness.

Earl was already a constant public figure in Woods's life, dropping his walking stick and trotting out to the final green at Sawgrass after that win for a long hug with his son—similar to the iconic moment between the two after the 1997 Masters. But both Tiger and Earl were deeply stubborn, and nothing put a larger rift between them than Earl's constant infidelity. According to Tom Callahan's definitive book, *His Father's Son*, Woods once cried to his high school girlfriend about his father's constant cheating. Earl and Tida moved into separate homes, and what kept them together was Tiger's rising career. In an infamous story from the 2001 Open Championship at Royal Lytham & St. Annes in England, Earl had an affair with a "cook." Callahan asked Earl about the woman, saying she must have been a good cook. "She sure knows how to keep that potato chip bowl filled up," Earl said. Supposedly a stream of escorts went into Earl's room at an event in South Africa, and Tiger once stopped talking to him after having to pay off one of his mistresses.

Tida would tell Tiger to mend the relationship, as Earl had suffered a series of heart attacks and wasn't likely to live long. By the winter of 2005, Earl almost constantly needed an oxygen tank. Woods spent part of that year's off-season break sleeping on the floor next to his father on Teakwood Street, and when Earl woke up on December 25, he threw a shoe at his sleeping son and told him, "Merry Christmas."

Woods was set to open his 2006 season a few weeks later at Torrey Pines, on the cliffs in La Jolla, another public course he grew up playing. (Once asked about his dream foursome and location, Woods said, "Just me and pops, at Torrey.") Three days before the event, he had one more of his countless tours of military facilities, this one being the Coronado BUD/S compound (Basic Underwater Demolition/SEAL training). Here recruits turn into Navy SEALs, renowned as the most difficult military training in the world. Woods spoke to a class and admitted that when he was younger, he had wanted to be a SEAL. He said if not for golf, he would be one of them, and that drew an eye roll from the instructors. After the tour, Woods approached one of the leaders, Thom Shea, who would earn a Silver Star while serving in Afghanistan. According to an article from Wright Thompson of

ESPN, Woods asked Shea how he kept his home life together with all the stress and travel. Shea would remember this clearly once all the messiness of Woods's personal life came to light. "You just do it," he told Woods. "You keep practicing."

In early May of 2006, Woods got the call he knew was coming but never wanted to get. He hopped in his private Gulfstream IV and flew with his wife, Elin, back to Cypress. In a closet in Woods's childhood room, an Obi-Wan Kenobi poster still hung. In Earl's bedroom, Woods's half-sister Royce was sitting next to their father, stroking his back and waiting for him to wake up.

Earl Woods was dead, and his loss left a void that Tiger could never fill—not with golf, not with the military, not with women, not with pain pills. He had been chasing perfection for so long that any sense of balance had disappeared. Buddhism had become a memory.


Golf was always serious to Woods, but things turned a corner in the summer of 1993 when Woods was seventeen. He had just lost in the semifinal round of the U.S. Amateur in Houston, dropping a match 2-and-1 to Englishman Paul Page, who will likely tell that story until the day he dies. That was the last match Tiger would ever lose in the U.S. Amateur.

After the defeat, Tiger and Earl left for Las Vegas to find the Butch Harmon School of Golf.

Butch's given name is Claude Harmon Jr., and he's the son of famed Winged Foot pro and 1948 Masters winner Claude Harmon. Butch had a weathered face, a paunch, and tanned forearms like a bricklayer. He had worked most famously with Greg Norman, but also with the likes of Davis Love III and Steve Elkington, players envied not just for their results, but for the beauty and effectiveness of their golf swings. Earl had taken a teenage Tiger on a world tour of the best teachers, considering it a privilege for them to have a chance to work with his son. Earl had also begun courting corporate sponsors, including Nike and Titleist, preparing for that inevitable moment when Tiger would turn pro and they would both cash in on all the hard work.

Earl had given Tiger his foundation in golf, but also knew that his gifted son quickly needed to move on. So at a young age, Tiger started working with two teachers from Southern California, Rudy Duran and John Anselmo. When Woods hit a growth spurt in his teens, Anselmo took Tiger's relatively flat and loose golf swing and made it more upright. It was to stop him from coming down with Hogan's most hated disease, "hookitis," as described by Earl in the introduction to Tiger's first instruction book, *How I Play Golf.*

That's how Woods showed up to the Harmon school—with a strong grip, a slightly shut clubface, all arms and legs swimming in baggy clothing, wearing sneakers and no glove. Holding the video camera to record that first session was Butch's son Claude III.

"It was pretty tough to impress my dad," Claude reflected, replaying the tape on a projection screen in an Orlando conference center. "And he saw this kid and he's pretty skinny. And you hear a lot about him but you never saw him up close. When he started hitting golf balls, it was so different than everybody else we had seen.

"He had Mizuno MP-17 irons, a TaylorMade Flex Twist driver with the black-and-white shaft. He used a Maxfli 90HT golf ball. And he could hit an 8-iron 200, 215."

Butch and Claude watched for a while, chatting with Earl, while Tiger continued to pound ball after ball into the dry desert air. This wasn't necessarily the first lesson, but more of a feeling-out period.

"What's your philosophy, how do you play golf?" Butch asked Tiger.

"I'll be one hundred percent honest with you," Claude III paraphrased as the young Tiger's response. "I swing as hard as I can, hit it as far as I can on every shot, and I have no idea where it's going. But

I hit it so much farther than everybody else that if I'm down in the trees, I know I'm going to be sixty, seventy, eighty yards past everyone else."

Butch asked Tiger how he would approach different scenarios of pin positions and wind directions. Tiger explained how he couldn't hit certain shots, especially the low ones. He wasn't yet strong enough in his arms and core to execute a swing path with an extremely descending blow and far left of the target while still keeping the face square—the type of attack angle that technology would prove to produce a straight, low shot. Not too much later, when he gained the strength, that would become the physical description of his signature 2-iron "stinger."

"But if you show me how to hit that shot," Tiger said, "I'll try and I'll figure it out."

As Earl remembered, one thing that Butch said stuck out above the rest. He told Tiger that Greg Norman had "a great set of hands, but you have the greatest hands I have ever seen." That got Woods's utmost attention. "I'll bet on the downswing," Butch said, "just before impact, you sense whether the club is opened or closed and you make an adjustment with your hands." Butch told him they would change his swing plane so that Woods could enter the impact zone more square, then use the talent in his hands to curve the ball in either direction. The face angle needed to produce a curve in either direction is no more than one or two degrees, open (for a fade) or closed (for a draw). Earl stepped away, and Butch and Tiger began working.

When Earl and Tiger had left, the Harmons went inside for a little palaver.

"What'd ya think?" Claude asked.

"I don't know if I'll have the opportunity," Claude remembered his dad saying, "but if I get a chance to work with this kid, I know I can make him the greatest golfer anyone has ever seen, because he has all of the natural tools that players need. He has such power and is so lithe and has such creativity. He could just do anything you asked him to do."

As for where Butch came upon his philosophy about the game, it was far more about history than it was about anything overly technical. Claude Sr. had been friendly with Hogan as they played events together on the hardscrabble PGA Tour of the 1930s and 40s. Claude Sr. listened when Hogan would talk about how he practiced and played, and those stories would be passed down to Butch and eventually to Tiger. When Butch didn't make it as a player himself, he naturally segued into being a coach. He bypassed the technical rhetoric that was starting to gain traction, directly rebuffing *The Golfing Machine*.

"I can't understand it, and I'm not sure a student at MIT could," he told *Sports Illustrated* when asked about the book in 2003.

He instilled that sense of natural feel into a young Woods. Butch didn't want such a talented player to be thinking too much about the machinations of every little movement. Butch wanted him to just play golf.

On the way to Woods's final U.S. Amateur victory in 1996 up in Oregon, Tiger turned to Earl and said, "I'm never flying coach again." He beat Steve Scott in an epic thirty-eight-hole final at Pumpkin Ridge and had a private charter waiting, set to take them to Milwaukee, where Woods would make his professional debut. On the Wednesday before the tournament, he held his "Hello, World" press conference, and after signing his $40 million contract with Nike, the company aired his famous black-and-white television commercial where he said, among other hyperbole, "There are still golf courses in the United States that I cannot play because of the color of my skin." He still needed Butch to float him a loan for the $100 entry fee, not having seen any of the Nike money yet. By finishing tied for sixtieth, he earned $2,544.

By the end of the year, Woods would win twice, be named Rookie of the Year, and be featured on the cover of *Sports Illustrated* as their Sportsman of the Year. The hype machine was running on overdrive, and Earl was feeding it at every turn.

"Tiger will do more than any other man in history to change the course of humanity," Earl said.

It sounded ridiculous, but Woods went out in his first major as a professional and blitzed the field at the 1997 Masters. After an opening nine holes of 4-over 40, he shot 6-under 30 on the back and finished the tournament at 18 under to set a scoring record at Augusta National, winning by twelve shots over Tom Kite. An estimated 44 million people tuned in to CBS for the final round on Sunday, numbering the largest audience for golf before or since.

Woods immediately became one of the biggest celebrities in sports, if not in the world. He changed more than just the face of golf, but its color and shape. He enthusiastically celebrated when he did something great in competition, just as they did in other sports. In a television interview with Curtis Strange, Woods famously said that he entered every tournament expecting to win. Strange, the two-time U.S. Open winner, laughed at him, but Woods didn't back down. It sounded like something Michael Jordan would have said.

Woods was beginning to make the game cool. He rebuffed previous golf norms, both in style and substance. He was more confident than he was humble, and the color of his skin was not lost on anyone—both for those who liked him and those who disliked him. His persona drew people in, and his overwhelming success on the course kept their interest. And his path to greatness was far from being technical or routine.

"I've learned to trust the subconscious," Woods said before that first Masters victory. "My instincts have never lied to me."

Yet less than a year later, Woods's instincts told him he needed to change. He was the Player of the Year on the PGA Tour in 1997, and he knew he still needed to be more consistent and versatile. He remembered what Butch had said back in 1993, and he wanted to implement a handful of changes in order to be more square coming into impact. He envisioned a drastic set of changes to his grip, setup, and swing path, and Butch warned that it could take a long time and could be painful. He recommended making one small fix at a time to ease the transition. Woods insisted on changing it all at once and also insisted on playing a full schedule of tournaments. So the 1998 season went by with one Tour win and zero majors, a disappointing performance for the young Woods. The naïve members of the peanut gallery began to call it a sophomore slump.

But Woods's instincts did not lead him awry. Once the swing changes settled in and felt natural, Woods went on a tear that was as good as golf has ever seen. He won twenty-one of the fifty-eight stroke-play tournaments he entered from 1999 to 2001, including five majors. His "Tiger Slam," which bridged the 2000 and 2001 seasons, was the first time anyone had ever held all four professional major titles. It was the closest anyone had come to Bobby Jones's Grand Slam of 1930, which included the U.S. and British Amateur titles, then considered majors.

Golfers the world over speak of that period in Tiger's game as if it were ripped from a fairy tale. The fellow touring pros talk about how the U.S. Open at Pebble Beach in 2000 was set up to be so difficult, the greens so hard and fast and the fairways as wide as bike lanes, framed by jungle-like rough, that a scratch golfer wouldn't have broken 100. Miguel Ángel Jiménez and Ernie Els tied for second at 3-over par. Tiger shot 12 under.

Instructors will speak about his perfect setup, his on-plane backswing, his unbelievably fast uncoiling of the hips while the arms and hands and clubhead all follow the path to the ball, dead square, as if any other path would have spun the earth off its axis. Any curvature of the ball was controlled with the subtlety of Woods's educated hands. Not even Nicklaus in his prime had his swing talked about with such high regard. Only Hogan was mentioned in the same breath of technical mastery. The link between Woods and Hogan was already being established, and it had been made through Butch.

Yet the difference was that Hogan came about in a time when instructors were hardly commonplace. Even Nicklaus and his life-long instructor, Jack Grout, had only one scheduled lesson a year—at Nicklaus's childhood home course, Scioto Country Club, in Ohio, sometime in the late winter or early spring when Nicklaus was getting ready for the season to start. The game back then was far more finite, the technical information limited and anecdotal. Before that, Hogan found the answers in the dirt because there was nowhere else to look.

That was not the case with Woods. He started to do his own re-search into physiology and kinetics, beginning a regimen of intensive strength training. Getting bigger and stronger also helped him draw closer to his obsession with the military. He no longer wanted to be a skinny, nerdy golfer. He wanted to be more like his friends in other sports. He considered himself an athlete and wanted to look like one. He always wanted to be the longest hitter on the Tour, and he had a trick of snapping his left knee at impact to get an extra ten to twenty yards when needed. By December of 2002, he needed surgery to re-move benign cysts in the knee and drain fluid inside and around the anterior cruciate ligament. When the surgeons went in, they realized that his ACL was only about 20 percent intact. More stress and abuse, and it would snap. The operation on his knee was the beginning of what would be a lifetime of physical ailments and surgeries.

The knee injury forced Woods to miss the season-opening tourna-ment in Hawaii for the first time in his pro career, but he returned ten weeks later to win the 2003 Buick Invitational at Torrey Pines.

"It's going to be week to week, trial and error, and see if I can keep playing," he said at Torrey Pines. "I know I'm not one hundred percent yet, but it's pretty darned close. It's a heck of a lot better than it was last year, let me tell you that."

Yet more damning was that Woods didn't feel as if he had anything to work on. No major overhaul of his swing was planned, no drastic changes were needed. He had no new military-like mission to focus on. Butch had taken an old-school approach and was there now for skills maintenance, not advancement. It was the same way Grout had dealt with Nicklaus. But the passive approach to coaching started to wear away at Woods. It certainly didn't help that his celebrity had risen to astronomical heights, and Butch hardly ever shied away from the cameras. Woods wanted someone in the foxhole with him, away from all the attention. That wasn't Butch.

In 1999, at the Dunhill Cup in St. Andrews, Scotland, Woods played a practice round with his good friend Mark O'Meara. Eighteen years Woods's senior, O'Meara struck up an easy friendship with Tiger, who even moved into Isleworth, the same private development outside Orlando where O'Meara lived. In Woods's down year of 1998, O'Meara won his only two majors—the Masters and the Open Championship—and collected the PGA Tour Player of the Year award.

"Here I am playing with this young, very talented, twenty-two-year-old individual who has actually kind of rekindled my spirit or my drive," O'Meara said in the summer of 1998. "And from the standpoint of my relationship with Tiger, I think that I am kind of a sounding board for him. I think he trusts me. We are good friends. He can confide in me. And in that way, that is how our friendship has developed, and I think it has helped him."

So at the Old Course in 1999, the two came to the famed eleventh hole, a short par-3 named the Eden, which normally plays into a left-to-right wind and has a green severely slopped from back left to front right. For a right-handed player such as Woods, the shot there is a low right-to-left draw with little spin, keeping it from ballooning into the breeze coming in off the Eden Estuary behind the green. O'Meara hit that shot with ease, but Tiger struggled with it. Walking beside the two men was O'Meara's teacher, Hank Haney. When O'Meara's shot bored through the wind, Tiger turned and looked at Haney, startling him when he asked how O'Meara hit that shot.

"I explained that it came from hitting a less lofted club and relaxing the arms into an abbreviated finish," Haney wrote in his 2012 tell-all book, *The Big Miss*. "He tried a shot, didn't like the results, and declared, 'No way I can do that.' But I knew he'd file it away."

Although Tiger went on to play outstanding golf through 2001, by 2002 his relationship with Butch was starting to sour. Tiger won the 2002 U.S. Open at Bethpage State Park on Long Island—the first U.S. Open to be played on a fully public golf course—but by that year's PGA Championship at Hazeltine, outside of Minneapolis, Tiger shooed Butch off the driving range and told him he'd like to work alone. Just before the 2003 U.S. Open at Olympia Fields, Tiger went for one last trip out to Vegas. When he finished tied for twentieth, that was the end of their decade-long working relationship. For the first year since 1998, Tiger had not won a major, and he fired Butch.

Early in 2004, after about six months without a full-time teacher, Woods and O'Meara went to play for big money in the Dubai Desert Classic in the United Arab Emirates, which became O'Meara's first win in six years. After coming in a tie for fifth, Tiger waited for O'Meara behind the eighteenth green, something Woods almost never did, and congratulated his friend. The two flew home together in Tiger's rented Gulfstream G550—he would soon own one for himself—and during the flight O'Meara told Tiger he needed to start working with a teacher again. Attempting to convince the best player in the world, with a world-class ego, that he needed to start working on his game was a tough conversation. But O'Meara had garnered enough respect from Woods over the years that the message began to sink in.

Tiger always had the innate understanding that what he feels as a player is not always what is happening. He had a saying that he knew he was playing his best when "feel" met "real." He understood proprioceptive dysfunction. He knew that no matter how well you understood your own body, and your own golf swing, it could always betray you. The smallest step off line can lead to troubling compensations, and those compensations can quickly begin to feel natural. If Woods acted immaturely off the golf course, he always had a prescient maturity on it, knowing that above all else to be alone is to be vulnerable.

So O'Meara suggested Haney, and Tiger remembered that practice round in Scotland. He liked some of Haney's swing ideas and liked even more that Haney was a Hogan worshipper. When the plane landed in Orlando, O'Meara's agent, Peter Malik, told Haney to expect a call. The next day, March 8, 2004, Haney was with his father at a steak house in Plano, Texas, when his phone rang. He didn't recognize the number, with a 407 area code from the Orlando region. He went outside to pick it up, spoke to Tiger for a few minutes, and when he returned to the table, he had just been handed the job that would define his career.

When Haney took over as Tiger's coach, Butch called him with one piece of advice.

"Coaching Tiger Woods," he said, "is harder than it looks."

The mountain of work that Haney placed in front of Woods was another one built on history, which Tiger liked very much. Haney wasn't promising perfection, but he was proposing a huge overhaul to get Tiger's swing closer to the way Hogan's looked. "There are very few perfect shots hit in golf, even by experts," Haney wrote. "It's above all a game of managing misses."

Haney had a tape of Hogan from 1964, when he played against Sam Snead in a great made-for-television match on *Shell's Wonderful World of Golf*. Both men were fifty-two years old, but swung the club as elegantly as they always had. Haney watched the tape often, including while he was a teacher at the sprawling eight-course Pinehurst Resort in the sandhills of North Carolina.

Haney was watching it again in the winter of 1981 when he had a little more wispy blond hair atop his wiry six-foot frame. With broad shoulders and a pronounced nose, Haney got up from his seat and began to draw lines on the screen with a marker. He found that the angle created by Hogan's shaft and the ground while at address was the solution he was looking for. That angle created a plane, the same plane on which Hogan brought the club back and through. It was different from what Hogan had described in his famous 1957 instructional book, *Five Lessons*. There, the swing plane was significantly more upright. It was illustrated in beautiful pencil drawings by artist Anthony Ravielli as being a pane of glass with a hole for one's head, and Hogan said that the club should never go over, or "break," the glass. Now Haney was seeing that the real plane, the one that the club traveled on, was the one created at address.

It was his epiphany, and when he met O'Meara at Pinehurst the following year, he found his first guinea pig for the theory, and good results followed.

When Woods called to hear what Haney had to say, he laid it all on the line. He wanted to get Tiger to bend over more at the hips, then "flatten" his path, attempting to keep the club "on plane" going back and coming through. A small squat with the legs as the transition began would also use the ground as leverage and create more clubhead speed. Unbeknownst to either of them at the time, it also created extra torque on Woods's lower back. The more bent over at the hips, the more difficult it is to maintain posture through impact—with a lot of the stress landing in the lower back, just above the tailbone. It can be managed by players with smaller frames, but with Woods increasing weight, especially in his upper body, the vertebrae were grinding away.

But Haney had a narrative, explaining that the ideology was based on Hogan. And it was found via the fabled method of trial and error.

It was the newest mission laid in front of Woods, and once he signed on, he attacked it with vigor. It took him a while to make the adjustments, and he didn't win a major in 2004. But in 2005 he won the Masters, then the Open Championship, and he was the best in the world again. He had gained weight, training intensely while layering muscles on top of muscles. Despite his incredibly thin waist, he looked more like an NFL running back than the kid who won the 1997 Masters. Tiger was a totally different person, in so many ways.

At the back of Haney's book is a list of Tiger's record from March of 2004 until May of 2010 stating that during their time together, Tiger won 39 of 105 stroke-play events, a 37 percent success rate. He also finished in the top ten an astounding 78 times, 74 percent. But most important, he won six major championships. The last was the 2008 U.S. Open at Torrey Pines, which Tiger won with two stress fractures in his left tibia, needing ninety-one holes to beat Rocco Mediate in a Monday playoff. With so much interest in the playoff, the stock market saw a dip in trading. Before the tournament, the people in Woods's inner camp—Haney, agent Mark Steinberg, and caddie Steve Williams—asked him if it was worth playing on his broken leg considering the potential aftereffects.

"I'm winning this tournament," Williams remembered Woods responding.

After that, Woods needed to take some time off to heal his leg. He and Elin had just had their first child, a daughter named Sam, and Elin was pregnant with their second child, a son they would name Charlie. They might have looked like the happy celebrity family on the cover of *People*, but even with Earl gone, Tiger was still his father's son.

Woods always had trouble sleeping and often took aids such as Ambien. Now with the injuries mounting, those started being mixed with pain pills such as Vicodin. Woods's infidelities became rampant, and they weren't all just flings. He became close with a New York nightclub hostess named Rachel Uchitel as they bonded over their grief—Tiger's father was dying while Uchitel's father had died of a cocaine overdose when she was fifteen and her fiancé had been killed in the World Trade Center on September 11. According to ESPN, she would often come down to Orlando with a friend, Tom Bitici, and Tiger would put them up in a condo. He would come over, shut the blinds, and the three of them would watch *Chelsea Lately*.

But Woods forgot to delete a text from her on Thanksgiving of 2009, and Elin picked up his phone and saw it. She texted back, posing as Tiger, then called Uchitel to let her know they were caught. The discovery was a long time coming, and Elin lost her cool, throwing the phone at Tiger and chipping his tooth. She physically attacked him, scratching his face before he could hide in the bathroom and text Uchitel back. Already dosed on a cocktail of pills, Woods left the house in a daze, getting into his Escalade and first driving it into a row of hedges, then another row of hedges across the street, then over the fire hydrant and eventually into a tree. Elin chased him in their golf cart, got to the scene of the accident, and broke the two back windows with a golf club.

When the cops finally got there, Woods was in and out of consciousness. He had been given a pillow and a blanket (likely from the neighbor whose front lawn he had crashed into) and was not entirely responsive. Elin said she tried to help him, but the cops never believed it.

And that book, *Get a Grip on Physics*, sat in the middle of the debris. What a strange footnote to such a monumental night.

When Woods got out of the hospital, he went to Mississippi, where he checked into a rehab facility presumably for "sex addiction." He didn't touch a golf club for well over a month, and in February, he gave his televised apology. When he thought about returning, he reached out to Haney, who had publicly said he was unaware of all that was going on in Woods's personal life. It was believable given how their relationship had cooled after the 2008 U.S. Open.

Woods had become frustrated with the way his game had evolved with Haney, and it was especially disconcerting when Woods's body began failing him. His search for answers led him to research not only physics, but also physiology and biomechanics. With the advent of 3-D motion analysis, the sequencing of the golf swing became more complete, along with the proper angular relationships of body parts. For example, the term *X-factor* started floating around Tour driving ranges, referring to the tension created between rotating shoulders and hips that led to a coil and release that brought about more club-head speed. Sophisticated equipment was also being used to measure the swing's stress on the body—especially the lower back—and physical trainers were working hand in hand with swing coaches to tailor each player's move to best suit the player's own physicality and limitations. Woods might still have been one of the best players in the world, but with Haney, Woods felt that the world was passing him by.

Woods planned a return to the game at the 2010 Masters, and the anticipation was enormous. On Wednesday of that week, Billy Payne used the annual public address from the Augusta National chairman to tear into Tiger, pulling no punches about how Payne felt concerning the events of the previous six months. "He forgot that with fame and fortune come responsibility, not invincibility," Payne said. "It is not simply the degree of his conduct that was so egregious but the fact that he disappointed all of us, and more importantly our kids and grandkids. Our hero did not live up to the expectations of the role model we saw for our children. Is there a way forward? I hope yes."

This absolute pressure cooker of an event was only moderately controlled by the strict policies of Augusta National. The excitement started to build when Woods shot an opening round of 3-under 68 and followed it with a 2-under 70 on Friday. It seemed this could be the most dramatic (and salacious) golf tournament of all time. After another 70 on Saturday in which he struggled on the greens and took 33 putts, Woods took a backhanded shot at Haney through the media: "I warmed up terrible. I didn't have control of the ball." After the interviews, he trudged straight to the range, the first time he'd done so after a round that week. When he saw Haney, Woods was steaming hot. "I hit it like shit," he told his teacher.

Haney was quiet, judging the touchy situation. "Tiger," he eventually said, "no, you didn't."

There was little talk after that, just a rushed practice session. With limited preparation time coming into the tournament, the two had decided on a conservative, but appropriate, strategy. Tiger was to hit mostly left-to-right fades off the tee, eliminating the dreaded quick hook. Taking away one side of the golf course—leaving all the misses to one side—was a classic strategy, one Nicklaus used for almost his whole career. Yet the limitations pushed Tiger over the edge. He always wanted to be able to hit every shot, which was why he went to Butch as a seventeen-year-old in 1993, why he and Butch changed his swing in 1998, and why he left Butch for Haney in 2004. But now he was cornered, and he hated it.

By Sunday morning, when Woods woke up just four shots off the lead, he vented to Haney, "I can't hit a draw," forgetting his score and fretting over his inability to move the ball right to left, a shot that many think is necessary for the right-handed player to win at Augusta. Haney was taken aback, explaining they weren't there to work on hitting a draw. Tiger wasn't happy, and it was clear.

A little farther down that range was the teacher Sean Foley. He had two students in the field, Sean O'Hair and Hunter Mahan, who had both made a name for themselves by striking the ball with consistent purity. While Tiger and Haney's practice was strained, Foley stood there with his camera bag slung diagonally across his chest. He had his own personal TrackMan, a new age ball–flight monitor that cost $25,000 and poured out numbers in explanation of each shot. O'Hair was at 1 under, eleven shots back of leader Lee Westwood, who would also eventually be under Foley's tutelage. Mahan was also in contention, six back of Westwood at 6 under.

Woods was just four back at 8 under, his knowledge from his previous four Masters victories proving invaluable. But he struggled as he kept trying to hit a draw, and Haney wasn't helping. Woods always knew how to get the best out of his game, scandal or no scandal, yet he understood that those two players down the range had an advantage. Tiger had played enough practice rounds with Foley's students that he knew Foley was operating on a more complex level, on a level that was not anecdotal, but scientific. Foley relied on the laws of ball flight, not the vision of ball flight; the laws of biomechanics, not the inconsistent history of previous success. Haney might have channeled Hogan, but Foley explained it with numbers.

Woods shot a final-round 69 and finished tied for fourth, five shots back of winner Phil Mickelson. Weeks later, Woods and Haney decided to split. Haney sent him a text message saying in part, "I can't tell you how grateful I am for the opportunity, but it's time for you to find another coach."

With his life in flux, Woods needed stability. He needed help, but he couldn't ask. Not yet.

He dragged a grim and lonesome face around for a while, his immense talent and grinding mentality inevitably getting him into contention from time to time. At that year's U.S. Open at Pebble Beach, a decade after his historic 15-shot win at the same seaside California course, he clawed his way near the lead before a Sunday 75 left him tied for fourth. He used to thrive on pressure, and now he was crumbling. He was no longer the same Tiger Woods. He would never be.

Three weeks after the Open, he went to Ohio, to the no-cut event at Firestone Country Club, where he had won seven of the previous ten years. This time, he finished his four rounds at 18-over par, beating just one player in the field, Henrik Stenson. Woods's total of 298 was 30 shots behind the winner, that same Hunter Mahan whom Foley taught, and 39 shots behind Woods's own total from 2000, when he won the tournament at 21-under par.

"Sometimes when things are going bad, it goes from bad to worse real quick, even for the best player in golf," Haney said that next Monday over the phone from his home in Dallas. "How all of sudden is his technique so bad? Now he's struggling and his swing is so bad? That part just doesn't make any sense."

Tiger thought differently. That Saturday night in Ohio, before the tournament was over, he picked up the phone and called O'Hair. He asked for Sean Foley's phone number. Woods needed something to hold on to, and he reached for science.

The ringtone coming from the camera bag on the floor was the voice of Big Daddy Kane, "Ain't No Half-Steppin'." It was August 7, 2010, and Foley would soon change the ringtone to "Paid in Full" by Eric B. and Rakim, and then to Mos Def. Foley looked down and saw an unknown number with a 407 area code, but he had had a feeling this call was coming and knew who was on the line.

Foley and his wife, Kate, had one child at the time, a two-year-old boy named Quinn. It was about nine o'clock on a Saturday night, and Quinn was sound asleep in the other room. Sean and Kate were sitting on a couch in the living room of a condo owned by Kate's parents in Burlington, Ontario, all four of them gathered around and relaxing.

Woods asked if he could pepper Foley with a few questions, and Foley said he just needed to put his son to sleep. Foley hung up, having lied to the most talented player in the history of the game about why he would have to call him back. Foley got off the couch, went to the kitchen, opened the refrigerator, and grabbed another Heineken. "Oh, yeah, dude, always Heineken." He told his family who had just called, gathered himself, grabbed his phone, and went out the door. He walked over the small porch and sat on a rock in the middle of a small adjacent park.

From there, he sipped his beer and called Tiger Woods back. When he came back into the house, his Heineken was empty. Kate and her mother, Randy, were eager to know how the call had gone. "Just how I had pictured it going for the last ten years," Foley said, smiling as he walked to the kitchen for another beer.

Foley is incredibly intellectually curious, researching anything he thinks is worthwhile. At five feet seven and 160 pounds, he wears tailored clothes (a tradition handed down from another Canadian golf icon, Ben Kern), with collared shirts normally buttoned to his neck. He always has a new pair of hip-looking eyeglasses. Growing up, he endured what he called "a childhood of cystic acne" while moving around all of North America with his family, eventually settling in suburban Toronto. In college, he played golf at the predominantly black school, Tennessee State University, in Nashville, where he did far more partying than putting up good scores. His favorite book is *The Autobiography of Malcolm X*, and he can converse about the evolution of the prefrontal cortex and its anthropological implications. He likes to say that his profession is a study of all that is around him, about the world at large. Golf is just a small personal puzzle, and he is trying to solve it.

"The mystery I don't like," Foley said. "I'm okay with things I don't know, but things I should know, I don't like that."

His obsession with being a golf instructor started at a young age, and he remembers being eighteen years old when the 1992 Canadian Open came to nearby Glen Abbey Golf Club, in Oakville. Foley went as an observer with his father, Gerry, an immigrant from the hardscrabble side of Glasgow, Scotland. From behind the ropes at the driving range, Sean pointed to David Leadbetter working with Nick Faldo, and Butch Harmon working with Greg Norman.

"*That's* what I want to do," Sean remembered saying.

In retrospect, he wasn't sure his father knew what he was talking about. Likely, Gerry thought his son pointed to Faldo and Norman, the two best players in the field. But Sean was pointing to their teachers, a position that seemed a lot more interesting than being a player. Sean Foley's mind ran wild with the idea of helping someone to make history. He started imagining himself on driving ranges at PGA Tour events, arm in arm with the best players in the world, telling them what to do. He thought about how much those teachers must know to have the trust of the best, and what an affirmation of knowledge it would be to have that trust. Foley thought there couldn't be a better job in the world.

He also remembered being at the 2000 Canadian Open, working as an instructor at Glen Abbey. Woods came through that week like a whirlwind, his magnetism drawing all the pasty Canadians toward him at every turn. (He also drew the eyes of Michael Murphy, on his couch back in California.) Over tequila shots at Philthy McNasty's (the actual name of a nearby bar), or while working as a waiter at The Keg chain restaurant when not teaching on the range, Foley would tell everyone who would listen that he was going to coach Woods one day. They all laughed.

But Foley kept putting in the hours and kept learning. He formed a relationship with Dr. Craig Davies, who opened Foley's eyes to biomechanical training. In their first meeting, Foley told Davies that he had recently lost fifty pounds on his max bench press and didn't know why. Davies put him through a couple of short exercises meant to "activate" the pectoral muscles, and Foley's full weight went up again.

"That's when I thought, 'What's this guy all about?'" Foley said.

The two would go back and forth for a whole summer, Davies

learning golf and Foley learning about the body. Davies would turn out to be the foremost physical trainer to PGA Tour players, while Foley wound up with "nightmare reflections" of how many juniors he had taught complaining about back pain because he didn't know exactly what was happening to the body.

"He approaches everything in a deconstruction type of way," Kate said about her husband. "He needs to break things down and find out how they work, or he's not satisfied."

By the winter of 2006, Foley joined his old friend Tom Jackson at Orange County National outside Orlando to run a small golf school aimed at talented teenagers. The market for this type of full-time academy was thriving, as places run by big-name teachers such as Haney and Leadbetter were drawing kids internationally, along with massive revenues. For the Foleys to move their family to Florida was a big jump, and Kate had to quit her job as a well-regarded IT salesperson in Toronto. But Sean continued to talk about how he was going to coach Tiger Woods one day. He thought it was his destiny.

"When I moved to Orlando," Foley said, "I didn't move two and a half miles from Isleworth by accident."

Like everyone else, Foley had watched Woods with a fervor, before and after the scandal. When Woods returned, Foley didn't just suspect that something was wrong with Tiger's game—he knew. And he could prove it.

"I'm just a huge fan, a total fan," Foley would say about how he felt on that first night when Tiger called. "But watching him win, like, every week, just hitting it dead sideways. Which is the truth."

Woods asked Foley to meet him a few days later at the 2010 PGA Championship at Whistling Straits, a bulging and beautiful golf course on the Wisconsin shores of Lake Michigan. Working on the range early in the week, Foley held a club against Woods's head to make sure it wouldn't move, and when he wasn't there to do it, Woods's soon-to-be-former caddie, Steve Williams, did it instead.

Seeing Woods work with a new teacher, no matter how informally, set the golf media atwitter.

Hogan moved his head (at least early in his career). So did Nicklaus. But Foley had biomechanical reasoning to keep the head still, the body rotating around the stable spine for a more repeatable motion. When Foley had more time, he also had Woods keep his weight a little more centered, meaning a little more toward his front (left) leg. He would use that leg to leverage the ground even more, a small dip with a bend in the knees enabling Woods to swing the club faster. Hogan did that later in life, and an extreme version of it was also the basis for a new swing fad at the time called Stack and Tilt, developed by two of Foley's friends, Mike Bennett and Andy Plummer.

But all of it was for a purpose. When Foley would eventually get Tiger on the TrackMan, the numbers were there. Just the way they wanted them. Just the way science told them was the proper way to strike a golf ball. It was finally quantifiable.

With so few close friends, Woods also confided in Foley. They both thrived on self-discovery, and they fed off each other's curiosity. They talked sports. They talked children. They talked far outside the realm of golf. Woods started to see his life in perspective again. The game was just their little piece of the world where they had moderate control. They worked hard to get some answers, and when there were more questions—which there always were—they would engage each other and try to find those answers together, through science.

But Woods was not the same person that he once was, and his body began to break down in extreme ways. His life off the course was hectic; he had to plan his playing schedule while figuring out with Elin when he could see their children. He spoke glowingly about fatherhood, but noted that he could no longer spend all day on the driving range or putting green.

In 2013, Woods won five times and collected his eleventh PGA Tour Player of the Year award, returning to his No. 1 world ranking. With reams of data collected by a new PGA Tour system called ShotLink, it became indisputable that Woods was again the best ball-striker in the world. Yet he had still not won a major championship since 2008.

For all of Foley's knowledge of biomechanics, he couldn't keep Woods's back from acting up. In August of 2013, Woods fell to his knees after hitting a wayward 3-wood during The Barclays tournament in New Jersey. He was suffering from back spasms, but still managed to finish runner-up. In March of 2014, he withdrew in the final round of the Honda Classic in Florida, shooting a 40 on the front nine and again citing the same back spasms. In April, he missed the Masters for the first time in his pro career, announcing that he had had his first microdiscectomy surgery on his back to relieve a pinched nerve. In early August, he withdrew from the WGC-Bridgestone Invitational in Ohio after he attempted to hit his ball out of an awkward lie in a bunker and it led to more back pain. The year 2014 was by far the worst season of Woods's career.

"I tried as hard as I could," Woods said after missing the cut at the PGA Championship in August. "That's about all I got."

Later that month, while Woods was at home recovering, Foley was at Ridgewood Country Club in New Jersey with some of his other students who were preparing to play in the first event of the FedEx Cup playoffs. The Tuesday before the tournament started, Foley was casually asked about Woods and gave a candid answer under the guise of generality.

"Sometimes, man," he said, "guys don't listen to what you're saying."

A few days later, his phone rang. It wasn't Big Daddy Kane, or Eric B. and Rakim, or Mos Def. It was a new phone, so the ring "sounded like I was getting eight million text messages at once," Foley said.

The number calling was already saved as a contact, but he didn't need to look down. "I knew who it was and I knew what it was about," he said. "And it was a very peaceful feeling."

On November 22, 2014, Tiger posted on his Twitter account that he had hired thirty-seven-year-old Chris Como as a "swing consultant." When Foley was once a guest on the *Charlie Rose* show, he mentioned Como as a rising star in golf instruction, and when reached for comment after the Como hiring, Foley said, "Yeah, that makes sense."

Como was never a high-level competitive player, but learned how to teach the game by going to school, taking lessons from guys such as Foley over the years, and storing all the information away. He was introduced to Tiger through Notah Begay, Tiger's old buddy from Stanford. Como was said to love TrackMan, and also loved talking about how the body understands and develops feel. Although Como made himself almost entirely unavailable to the media outside of his relationship with *Golf Digest* and Golf Channel, Woods had seemingly found someone he could bounce ideas off of rather than someone who would only give him direct instruction.

"I had this plan in my head of where I wanted to go and what I want my swing to look like and what I want to get out of my body and out of my game," Woods said. "I just needed to align myself with a person that felt the same way. Chris fits that for sure."

Foley showed no signs of bitterness and said, "Tiger is my friend and has remained so. There's no one pulling for him harder than me."

After his disastrous 2014 season and his breakup with Foley, Woods started 2015 with his worst score as a professional, shooting a third-round 85 at Nicklaus's Memorial Tournament in June, in addition to his worst finish in a major, shooting 80-76 at the U.S. Open at the links-style Chambers Bay on the shores of the Puget Sound.

"I've had people come up to me and say, 'Well, this really validates what you did,'" Foley said. "And I'm just like, 'Why do you even think that?' To me, I can't imagine ever thinking like that. I don't think it validates anything.

"It just shows that, you know, I think if you asked a lot of people who have climbed Mount Everest, if you interviewed them and asked, 'If you knew what it was going to be like, would you do it again?,' I don't think many of them would say yes."

In April of 2017, Woods had his fourth back surgery in three years. At 3:00 a.m. that Memorial Day, he was arrested in Florida on charges of DUI. He was severely intoxicated on pain and sleeping pills, and his infamous mug shot and the footage of his stumbling arrest would become Internet fodder forever. He went away to another rehab facility and once again issued a public apology.

"I would like to apologize with all my heart to my family, friends and the fans," he said. "I expect more from myself too. I will do everything in my power to ensure this never happens again."

There was no mention of Buddhism, and any sense of balance that had temporarily come back had seemingly disappeared. Woods's constant search for something more tangible led him to the cutting edge of science, and it was still never enough. He could never separate obsessive striving and happiness, nor complacency and unhappiness.

Woods and Como split just before the start of the 2018 season, and Woods did seem to have a significant change in attitude. He was more approachable. He had made friends with the younger players on the Tour such as Justin Thomas, Patrick Reed, and Rickie Fowler. Woods showed patience with his game, and poise on the course. And he started to play a lot better.

He almost won in Tampa in March, finishing second, and shot a blistering final-round 64 at the PGA Championship at Bellerive in St. Louis, putting a charge through the tournament but coming up two shots short of winner Brooks Koepka. Woods earned a spot in the FedEx Cup playoffs for the first time since 2013 and made it all the way to the final thirty golfers that qualified for the Tour Championship. At East Lake, where Bobby Jones grew up fishing in the pond, Woods reentered the winner's circle with a resounding victory that transcended the sport. It continued when he won the Masters the following spring, his fifteenth major championship and first since 2008. Woods had reinvented himself as a player and a person, without a full-time coach.

"I think the fact is, there is a truth that we operate from as people," Foley had said, "and acceptance of all the external factors is key to joy."

After that second phone call that ended his professional relationship with Woods, Foley took his two sons to school. He went to the gym, then went home to hang out with Kate. He found more time to work with some new players and set up a relationship with an indoor studio in the middle of Manhattan known as Golf & Body NYC, where he hopes to keep a presence as his own life segues into something a little more quiet and stable.

"I don't really ask 'Why?' or 'What if?' very often," he said. "I just don't think they seem to lead to any real answers. They lead to fictitious story lines where we think we're figuring it out, and we're actually getting more lost."

# FOUR

# SCIENCE IN SLIDELL, LOUISIANA

Corey Weber picked up the phone in the pro shop at Pinewood Country Club, a small, cramped room with AstroTurf carpeting and a low, hanging ceiling. Weber was an assistant golf pro here, spending as much time as he could away from the register, reading and studying and listening to his mentor, who was standing ten feet away, a broad tree among the racks of shinny clubs. Large and lumbering, James Leitz is six feet three, and through his frameless glasses, he struck an unmistakable presence, loud and warm and above all Cajun—as irreverent and funny as he is serious and loving.

Leitz was five minutes late for an early-evening lesson, not something entirely uncommon. He stops and talks to anyone about anything, never disregarding an opportunity for connection, always in a rush but never too late. A single question easily turned into a forty-five-minute conversation about the hot Louisiana weather, about how the cooking of his northern-Italian mother from Pittsburgh wonderfully evolved to incorporate the creole, about how his father moved the family from his native city of New Orleans to here, Slidell, on the north shore of Lake Pontchartrain, to avoid the drugs and mischief that began creeping up near their house under the highway. That was where a lot of Leitz's childhood friends still lived, if they weren't dead.

But him, Mr. Smartass in the classroom, Mr. Hotshot on the golf course, he landed at this blue-collar country club where he began thirty-one years prior, raking bunkers and fixing divots. It was now March 31, 2011, two days after the twenty-ninth birthday of his son Ross, the oldest of his three boys from a marriage that dissolved, what he would describe only as "a tragedy of Hurricane Katrina." Leitz was heading up the small hill to his hand-built garage on the range, large enough for two cars but instead packed wall to wall with technology, computer monitors, and a white faceless mannequin wearing a scant black vest and knee braces dotted with electrical nodes. The clock in the pro shop above the window with the brown tin blinds said 5:06 p.m., and James had started toward the door when the phone rang.

"Sean Foley?" Weber asked in his New Orleans twang, totally disbelieving, the name stopping James dead in his tracks. "Tiger Woods's teacher?"

Weber put his hand over the mouthpiece and looked at the doorway. "James," he said quietly, holding the phone away from his face, "this is Sean Foley on the phone."

The two men, mentor and pupil, shared a moment of silence. Neither moved for what seemed like an eternity. Rarely was it ever quiet around Leitz, but this was not just another moment. This was the moment when things were about to change, when decades of work, more futile than fruitful, would come to a head. Leitz never knew if things were going to work out for him, because they rarely had.

But now was not the time to waver. Leitz broke into a smile that split his jovial, round face. He told Weber to take Foley's number, he would get back to him after the lesson. When destiny finally came calling, Leitz knew it would be inappropriate to give it short shrift.

The garage where Leitz works sits at the left side of the grass range, its green siding facing the gravel parking lot and the ranch-style clubhouse. From sunup to sundown, the tan garage door is almost always retracted into the ceiling, opening up this small science lab to the outdoor world. Leitz built the garage in 1998 to the tune of about $20,000, the investment necessary to create a space for private lessons out of direct sunlight and ensure that if someone drove a hundred miles to tap into his knowledge, the rain could not get in the way.

Leitz tumbles into the room like a wrecking ball, wearing a collared black shirt buttoned to the top with TITLEIST on the left breast, flowing gray trousers, and black leather golf shoes. At the front of the room, closest to the open garage door, are two plastic mats kept together by a small wood frame, another thing Leitz built by hand. Near the far wall are five large staff golf bags, all filled with an assortment of clubs, mostly old. The wall is covered with framed photos, signed golf flags, and pennants from such places as MIT. Leitz rarely walks past the wall without looking at it. The assemblage of knowledge and experience is a testament to his hard work, and to the consistency of physics. That display means the world to him.

Behind the hitting area, he plops down in his high-backed black chair, the faux leather and cheap plastic squealing under his substantial weight. He perches behind a desk of sorts, an L-shaped piece of black furniture that is hardly discernible under the tangle of technology that sits atop it. A laptop computer is swimming in the middle, facing a separate black keyboard. To the right, a thirty-five-inch monitor sits on the desk, topped by a hanging twelve-inch monitor protruding from the wall. On the right arm of the desk is a dirty beige keyboard, and above it, recessed in a brown wood bookshelf with no shelves, is a twenty-inch monitor, and above that another thirty-inch monitor, suspended. That's four monitors total. Somehow, near the left of the desk, a black mouse rests on a blue mouse pad, and behind it an iPhone is propped up against a notebook, next to a stage microphone lying on its side.

The screens of all four monitors are almost always on, and almost always showing the same thing. Directly behind the hitting area is a video camera on a tripod facing down the proposed target line, and the motion it captures is immediately displayed on the left half of the screens. There is the slightest delay, maybe one second, so watching someone casually walk past is like seeing life on replay. In this temple to the objective, reality is a sequence of moments. There is no mystery, no guessing, no such thing as the indescribable. Here, life is precisely represented by machines and physics and geometry. There are no lies, and the truth is plastered in pixels on the screens and in the psyches of those that walk out. "My job is to get you to hit the shot the way you want to!" Leitz exclaims, which is how he says almost everything. "It's like crack cocaine—you'll keep coming back!"

Underneath his desk are two surge protectors packed to capacity, and the desktop computer where he will burn a DVD of both the audio and video from every lesson to give to his students on the way out. Also down there, pressed against the left corner closest to the mats, is the holiest of holies. The TrackMan is hidden from plain sight as if to give the impression that the numbers and graphs filling out every monitor screen come from some foreign intelligence that only Leitz can conjure. Yet when the curtain is pulled back, the real intelligence is Leitz's ability to decipher those numbers and curved lines and explain them in plain language. That is his talent, his gift, the reason people come from all over to see him, and the reason Sean Foley called him. The satisfaction of understanding the physics and geometry in action is why Leitz so passionately loves his work, and giving him the data he needs, the data he strived so long and hard to obtain, is this little machine hidden under his desk.

"If someone hit it," he said, "I'd cry."

When Foley was riding at the height of his popularity as Woods's coach, he and Leitz gave a presentation to the Illinois PGA at a sleek convention hall outside Chicago. Foley was the main attraction, and Leitz's presence made the presentation feel as if the two were Abbott and Costello with wireless microphones. Leitz played the straight man, speaking dense geometry with interjected creole analogies, while Foley waxed poetic about the power of this information and how he used it in such situations as when standing behind Tiger Woods on a driving range and coming up with something smart to say. Most of the pros in the audience zoned out as Leitz talked through his PowerPoint projections of graphs and numbers, but when Foley said the word "Tiger," they all snapped back to attention. That's the way it went for the two coaches on their little tour dates, and they became one of the most sought-after acts in this strange and self-sustaining genre of scientific golf instruction.

As a teacher, Leitz had a unique, modern approach. But just as *The Golfing Machine* had few certified instructors and was used more as a tool, the same went for people such as Leitz, who became most useful as an information resource. Some counterintuitive geometrical concepts had been proven with the advancement of ball-flight monitors such as TrackMan, and the same can be said for the three-dimensional motion analysis provided by the vest and accoutrements hanging on the dummy in the Leitz's garage studio. These concepts were not easily understood, and Leitz was at his best in breaking down such dense ideas to make them relatable. He drew a certain type of student, and many of them were fellow teaching professionals such as Foley. Not a lot of recreational golfers (or touring pros) have the time, energy, or capacity to internalize all of the technical minutiae in the work of Leitz or Homer Kelley. The people who are drawn to this type of information harbor a deep intellectual curiosity. They are like-minded people who seek answers to their questions about why things happen. Once the why and the how are explained and understood, then improvement should be an easy next step. In this world, knowledge is all the power needed to implement change.

To be close to that knowledge was to be close to the power of the game, so people were drawn to Leitz. Foley took most of the questions at that presentation in Illinois, answering countless inquiries about his work with Tiger. But Foley often brought Leitz into the conversation, giving further insight into the scientific explanations.

The most important concepts that had recently been cracked were the laws of ball flight. Conventional wisdom said that the ball began on a line that was dictated by the path of the club. Turns out that about 90 percent of the ball's initial route is actually determined by the direction the clubface is pointing. Foley likes to tell a story about how when he was a teenager at the Canadian Open, he would set up camp at the short par-3 fifteenth at Glen Abbey. With almost all right-handed players, the divots were aimed left of the green, yet almost every shot went in the direction of the flag. *Why?* Now, Leitz explained it with numbers that described the geometry.

"When you hit a ball with a descending blow," Leitz said, "a path that looks to be aimed left of the target line is actually square at the moment of contact. Every good wedge shot should leave a divot looking left of the target." Foley was smiling, and that got the pros in the audience to start taking notes.

Turns out the inverse is true, as well. When hitting a driver, the ideal launch conditions for an average swing speed call for an ascending blow. Square to the target line while moving up means a path that is from the inside out, or aimed to the right for the right-handed player. In the parlance of TrackMan, that means the attack angle on a short iron should be somewhere around minus four degrees with a path of about minus two, while the driver is best around plus two and plus two, respectively. Aiming to hit in the center of the face (creating a high "smash factor") also sets the proper launch angle and spin rate for optimal distance and control.

Recalling and applying these concepts while trying to swing is difficult. Foley dealt with this challenge daily with his students, and Leitz had his own way of imparting the wisdom of geometry. Back in Louisiana, just weeks before that presentation, he had poured himself into his two-door, manual-transmission sports car and driven to a local restaurant for lunch. He began explaining all these concepts and how their discovery blew him away. In between exclaiming his love of physics, he was greeting and chatting with everyone who passed by. Before he could even order, the waitress came over and placed crabmeat au gratin, red beans and rice, and a plate piled high with fried oysters, fried shrimp, fried catfish strips, fried potato hush puppies, and two slices of buttered toast on the red-and-white-checkered plastic tablecloth. In a flurry, it all disappeared.

Leitz returned to his teaching shack for another lesson that started late and ended even later. By four o'clock the shadows were lengthening, and in walked a man in his midfifties, carrying a small golf bag. Ron Hickman owns, operates, and works as the teaching professional at Timberton Golf Club, up in Hattiesburg, Mississippi, a little over an hour north of Slidell. About five and a half feet tall, Ron was casually dressed in olive Dockers and a floppy baseball hat, while his lingering staccato laugh was infectious as it filled the heavy early-evening air. Behind him walked his wife, Pippa, wearing a smart white golf skirt and collared tank top over a slight frame, her blond ponytail popping through the back of her white visor. James had called Ron and Pippa friends for a long time, so this meeting was less of a lesson than a rap session. Ron put down the golf bag, and Pippa grabbed her 7-iron to warm up.

Everyone was chatting and laughing, while the images of Pippa swinging in that strange delayed-live feed ran on all four screens. She hit one solid, flying low and maybe 125 yards, a touch left of target. A little mechanical "bing" sounded, and the screens went black for a quick moment before resuming with the recording paused to her setup and the TrackMan numbers surrounding. Ron was in the middle of telling Leitz a story about the 2010 PGA Championship at Whistling Straits in Wisconsin (the first time Foley and Woods worked together), where Ron was a rules official. Leitz couldn't help but glance over at a screen and take a peek at the numbers. He didn't say anything to Pippa and continued listening to Ron describing where he was when Dustin Johnson ground his club in that "sand trap" on the seventy-second hole.

After some time, Pippa pulled out her oversize driver and began hitting balls out into the beautifully purple sky. She didn't generate enough clubhead speed to hit shots far off line, but two consecutive shots flew low and right. Leitz looked up to the screen and saw that the last one flew 138 yards, with the club approaching the ball at a descending angle that was to the left of the target line. Pippa leaned on her club as James pulled out a white three-ring binder, flipped to a laminated page with a chart of numbers on it, dragged his finger over to a certain column, then down to a certain row, and lightly tapped the box he was looking for. It had the exact attack angle, swing path, launch angle, and spin rate that would maximize distance through carry and roll for Pippa's clubhead speed. He explained to the room that to maximize distance, she needed to hit up on her driver and out to right of the target line. Understanding that information was Leitz's proprietary knowledge, and his lessons were mostly his just translating it.

"So now, sweetie," he said as he tossed a ball from his seat behind the desk to the wood-framed mats, "go ahead."

Pippa teed it up, waggled just enough, and crushed one with a high, right-to-left flight. Leitz hardly smiled and looked at the screen, where it said that her exact same eighty-one-miles-per-hour clubhead speed produced a ball that now flew 174 yards.

"If my fly is open because I'm too fat," Leitz explained, "should I go out and get bigger pants, or should I just lose some weight? Hell, I gotta go out in the world!"

"Now that's Cajun psychology!" Ron cackled. "Do they have that in New York?"


Two days before Foley called, Leitz sat alone at his desk, thinking about a student, Rick Frieberg. Frieberg had recently traveled from South Florida for a lesson, returning home with the first piece of instructional equipment that Leitz had ever built. After spending over $10,000 for various patents, Leitz had developed a small mechanical unit that could explain many aspects of ball flight, including the mysterious D-Plane. The device had a lofted iron with a small plastic stick pointing out of its face that moved along a sheet of plywood in a single-pendulum motion, at the base of which on the ground was a four-inch-wide strip of wood with a bisecting yellow line, looking like a runway. The yellow line represented the target line, the stick out of the face represented the swing path, and the plywood represented the angle of the swing plane. Got it?

Leitz had explained it all to Frieberg and convinced him to buy the first model. Yet by the time Frieberg got back to Florida, he could no longer remember precisely how to use it. He called and asked Leitz to explain it to him again, and rather than just speak over the phone, Leitz said he would make a DVD and send it to him in the mail. To help other students remember, Leitz recorded each lesson from start to finish and let the student go home with it. "You only retain about ten percent of what you hear in the lesson," he said. "That doesn't seem right. You should be able to go home and get all of it." There in his office, Leitz hit RECORD and starting talking—no script, no notes, nothing.

In eleven minutes and thirty-six seconds of voice-over, Leitz used two example videos of his friend and fellow teacher Rob Noel, along with a clip of himself with a Hula-Hoop and a distant fluttering flagstick, and, of course, different recorded camera angles of his model in motion. He spoke clearly, slowly, and concisely, simplifying the theories as much as possible.

The basis of the segment was to dispel some myths about the golf swing that had come along with proliferation of video in the 1980s, mostly that the path of the club created a plane, and that plane was a main determinant of where the ball would go. (Again, the geometrical truth is that the direction of the face is the overwhelming determinant of the initial flight of the ball, not the path of the club.) "The path of the club is ever changing on the inclined plane," reads one headline on the video, superimposed in red text as the images play in slow motion—forward, then rewinding.

The video describes what golf physicists such as Leitz have dubbed the D-Plane. The mechanical description is, it's the angle created by (1) the path of the club in three dimensions, meaning both angle of attack and path relative to the target line; and (2) the position of the clubface at impact. Leitz has another model to explain this, where a comically oversize clubface the width of his broad shoulders has two plastic sticks coming out of it, both able to move in all directions. The bottom one is red and represents the path of the club; the top one is yellow and displays the direction of the clubface. In a vacuum, the ball spins on one axis only, and that axis is created by connecting the straight lines of these two directions. If the club is moving left and the face is square with the target line at impact, the axis will be tilted and the ball will spin its way into a left-to-right flight. That is the physical description of how to hit a straight fade—a shot that starts flying straight and then falls into a soft fade, the shot shape that Hogan preferred. Swing along a path that aims more left with a clubface closed to the target line but open to the path, and there is the pull slice.

(This is all assuming contact on the center of the face. If it's an off-center hit, something called the gear effect changes the way the ball spins. Hit it off the heel, and the ball will spin more right to left, while the inverse is true for shots off the toe. For all the marvels of TrackMan, no one has yet been able to calculate the point of contact with much precision. TrackMan's "smash factor" is ball speed divided by clubhead speed, which determines how much energy is transferred from club to ball. A total of 1.50 is great, but that's as close as Track-Man can get to showing how flush the shot was hit. A low smash factor shows the shot wasn't hit well, but the system can't say if it was off the toe or the heel, thin or fat. So Leitz goes to the local office-supply store and buys a case of Magic Markers, colors in the face of the club, then looks at the face to see where the ball left a mark to determine where contact took place.

"Twenty-five thousand for TrackMan," he says, "and a dollar twenty-five for everything else.")

In this corner of the golfing universe, the common saying is that the ball doesn't care who hit it. The flight is nothing more than an objective result of an applied physical force. Foley was once in a grillroom watching an instructor on the Golf Channel who said that a draw goes farther than a fade. Foley laughed in disgust. "Do you think the ball knows if you're a righty or lefty?" he asked. "You can put the same spin on the ball from either side. I don't even blame these guys. They just don't *know*."

But history's great players had an intrinsic knowledge of ball-flight laws, even if they couldn't describe exactly what was happening.

"The old guys, they closed their stance for the driver and opened it for irons," Leitz says, intellectually echoing Hogan, the man Leitz studied more than any other. "They didn't know why, the geometric reason why. Now we know, and that's the cool thing. It all connects."

Instead of burning his video tutorial on a disc and mailing it to Frieberg within in a week, Leitz decided to make it immediately accessible. He already had a YouTube account under the pseudonym ForeverMySaints, as the first video he ever posted was a ninety-six-second clip from an old cell phone camera capturing the New Orleans Saints' winning field goal in the 2010 NFC Championship Game. From his seat in the upper deck of the Superdome, Leitz recorded Garrett Hartley's winning kick, the ensuing chaos that surrounded him, and the big kiss he landed on his soon-to-be-ex-wife. In the description, Leitz wrote, "Video Shot by James Leitz, who as an 8 year old boy attended the first Saints game they ever Won in 1967. We have reached a goal we never thought was attainable. Everything from this point is Langiappe [sic] !!!! Go Saints!!"

Then, on his son Ross's birthday, Leitz posted the video titled "D Plane Model Video by James Leitz" and sent the link to Frieberg. Leitz took his son out for dinner, and the two ate crawfish and fried jalapeño hush puppies. By the next morning, after Frieberg had forwarded the link to some teaching pros he knew, the video had 70 views. By the following day, March 31, it had 170 views.

One of those views was from Foley, who then picked up the phone and called Pinewood, looking for the man who had made it. Less than two months later, after extensive conversations about the D-Plane and other geometrical phenomena, Foley was the subject of a profile in the *New York Times*. He was asked about accusations that he had stolen ideas, specifically from two teachers named Mike Bennett and Andy Plummer, whose approach was known as Stack and Tilt. Foley may enjoy attention, even crave it, but he was not one to pretend that everything he says and teaches is born of his own genius.

"Mike and Andy, they're buddies of mine," Foley told the newspaper. "We talk all the time. I'm on the range ten hours a day and I talk to everyone. How is this stealing? This is about growing the game of golf."

Then, fully cognizant of what he was doing and whom he was talking to, he said: "I've learned a lot, for example, from a guy named James Leitz, a hidden gem down in Louisiana."

When the story ran on May 10, 2011, Leitz's phone began ringing incessantly and his lesson book filled up months in advance. He had already given talks about golf and science all over the country, including at MIT, where he got that pennant to hang on the wall of his teaching studio. Now he was forced to turn down lessons. No longer was he a hidden gem, not after being name-dropped by Tiger Woods's current coach, the man who by that distinction alone was at the forefront of golf's instructional body. And at the front of Foley's mind was Leitz, an inspiration and an information resource.

"When that call came, obviously I was excited about it, because Sean was teaching Tiger," Leitz says. "But also, I've never been intimidated by anybody talking about the golf swing because it's something I've done my whole life."

Grappling tirelessly with what was once indescribable, Leitz finally hit upon conclusions that he could prove, conclusions he carries around the world like a gospel. To witness his creole preaching is like being in a class where voodoo spells are explained, and everyone laughs, eats, and goes out for a drink. He did not become this missionary by happy divine intervention, and he never once sat on his hands and waited for an epiphany. He walked into every dark corner and shouted for answers, often hearing nothing more than his own echo. Eventually, someone, something, responded, and the lights began to flicker on.

"Hard," he says. "Hard, as I tell people. I've worked it hard. And if there's somebody out there that's worked it harder, God bless them. But I feel like I belong."

Leitz was born to a helicopter mechanic named Jim and a stay-at-home mother named Pam in a house underneath Interstate 10, the only boy joining five girls. The family lived so close to Louis Armstrong International Airport that every window in the house was cracked from the percussion of overhead planes. Somehow, James slept through the shaking of the jet engines and played baseball, basketball, football, and ran track. With his big body helping him to excel as a center and a defensive end, the football coach at John Curtis High School recruited him as an eighth grader to join their powerhouse program, the best in New Orleans, a premier school with only about 150 students. The coach came to the tough areas of the city every so often to see what kind of young talent he could find, and in the game he watched, Leitz said he knocked out the opposing team's punter and quarterback on two separate plays.

Leitz was ready to play football for John Curtis when his father moved the family out of the crumbling neighborhood, where James was getting bored with middle school and his mischievous cutting up was pointing him in a downward direction. When his friends started getting involved in drugs and getting in trouble with the law, Jim went looking for a better life for his family and took them to the north shore of Lake Pontchartrain, to Lacombe, a rural enclave about forty minutes outside New Orleans proper.

"You go from a neighborhood with kids that you grew up doing everything with, and then you move," Leitz said. "We were the kind of neighborhood where nobody got transferred or promoted. The guy that drove a bread truck didn't get promoted. The guy that was a plumber's helper . . . you know what I'm saying? Everybody stayed, so when someone moved, it was tragic."

When the family landed in Lacombe, they were eleven miles west of Slidell, the small northern hub with about seventy thousand residents. No one else lived within a mile of their property, and for James the city kid, it was crushing.

Bored to tears, James looked in the back of his dad's beat-up Chevy and found a set of rusted old golf clubs. Back in the fifties, Jim had tried golf a couple times before hurting his back, and the clubs now sat idle. Growing up, Leitz never knew anyone who had played a single round of golf—including his dad, who had never mentioned these clubs—and never knew anyone who had gone to college for even a single day. But when he picked up the clubs late one summer afternoon and starting smacking a ball around the family's acre and a half of land, it took him less than an hour to have a realization.

*Wow . . . This. Is. Fun.*

Leitz never played a down of football at his new school, Slidell High, and never played another organized sport besides golf. He never had a formal lesson, but started reading all the publications and books he could get his hands on: every *Golf Digest* and *Golf Magazine*, plus both of Ben Hogan's books, *Power Golf* and *Five Lessons*. The school golf team at Slidell wasn't good, and despite being unable to break 80 in competition, James played No. 1. Occasionally he could shoot in the high 70s when he was on his own, but under the pressure of a tournament, he would blow up.

When he turned fifteen, Leitz went with his buddies to Lakewood Golf Club, just south of the city, to watch the pros at the New Orleans Open. It was 1974, and Jack Nicklaus was in his prime. Walking off the third tee, Nicklaus stopped, picked up his foot, and flung a piece of gum off the bottom of his spikes. Leitz scurried under the ropes and retrieved the gum, his friends laughing at his innocence and idolatry. Leitz brushed it off and brought the piece of gum home. There, he split the dried gum in two, putting one piece in his wallet and leaving the other piece in his room—just in case he lost the one he carried.

"Well," Pam told her son, "at least he didn't step in shit."

By the time he graduated high school in 1977, Leitz also had a patch of grass in his backyard that was very different from the rest. Growing as high as three feet tall was a small area of springy Bermuda grass, a combination of three divots taken by Nicklaus, Arnold Palmer, and Gary Player. That patch of grass was so cared for and nurtured that the encroaching crabgrass had been fought off. Leitz had collected Nicklaus's discarded piece of turf in New Orleans and gone to two separate tournaments in Florida to collect his commemorative pieces from the rest of the Big Three.

"No one loved the game more than me at that time in my life," he said.

There was no letup in Leitz, no question of what he wanted to do the rest of his life. He did not receive any scholarship offers to play golf in college, so he decided to take a year off school. He got a job at Pinewood as an eighteen-year-old willing to work all hours, as long as they'd let him play. So there he went, fixing divots, raking bunkers, planting seed. He did whatever grunt work was asked of him, then he practiced until the sun went down.

By the following fall, he applied and got into McNeese State, three and a half hours west of Slidell in Lake Charles, Louisiana, beyond Baton Rogue and Lafayette and every other city Leitz had reasonably expected to ever visit.

The golf team at McNeese had just been to the NCAA Tournament the previous spring, playing the event all the way up at the historic Course at Yale, in leafy New Haven, Connecticut. Leitz showed up and was lucky to walk on to a team that had only six players. His scores in the tryout fell in the mid-70s, and he found a spot on the end of the roster.

He continued to put in long hours on his game, but there was a big problem. He wasn't getting any better. If anything, he was getting worse.

"If you work harder than everybody," Pam told him, "you'll beat them all."

"Mom, you've never played golf."

Leitz was now a spindly six feet three, with shaggy hair past his ears just like Tom Watson. Leitz could hit the ball high into the sky, way over trees and obstacles, always thinking he could do this stuff because he was special. "I didn't know it was because my swing sucked," he says now.

After his freshman year, coach Hubert Boales—who was also the linebackers coach on the football team—sat Leitz down for a chat. Boales said that as a freshman, Leitz wasn't expected to contribute, but because he was a good kid, with a good attitude, he could stay on the squad. By his sophomore year, Leitz was playing even worse. Following that season, Boales sat him down again and again said he was keeping him on because he was a model of hard work for the rest of the team, but he needed to be mentally stronger.

"Of course you're going to choke," Leitz says, "because you have terrible mechanics. It's like someone shooting free throws over the back of his head and saying, 'Man, he's just not making them in this free-throw tournament.' Well, he's shooting it over his freaking head!"

After that talk with Boales sophomore year, Leitz was done with moral support. He wanted to contribute to the team, but all his hard work made no difference, and his frustration was boiling over. His proposed life's work was in jeopardy, and he hated it. No one could even point him *in the direction* of an answer, so he went about finding it himself.

He and his then girlfriend Gina bought a cheap camera and a roll of film with twenty-four pictures. They went to a field on campus and Leitz came up with a plan. He told Gina to take twelve pictures of him from behind, down the target line, and try to snap each one at a different moment in the swing. Then she should go face on and take another twelve, therefore using the whole roll as efficiently as possible.

She tried her best, and one agonizing week later they got the photos back. He laid them out on a table in sequential order and thought he knew what he was looking at. Remember, he had stockpiles of magazine articles and knew Hogan like the back of his own callused hand. From the face-on view, a tree was in the background just about on line with his head. He saw that on the downswing, his head dipped lower on the tree, closer to the ground, and figured that was the problem.

*Well, hell,* he thought, *I've just got to keep my head steady.*

What he didn't notice at the time was that the picture closest to contact showed the head of the club at the ball while his hands were closer to his right pocket. In teaching parlance, the club had "passed his hands" and had therefore added loft, resulting in all of his high shots, also making it difficult to achieve solid contact. Leitz keeps that image dear to his heart, plastered as the background on every one of his computer monitors in his technology-stuffed teaching shed.

But that knowledge would come later. Now, he had this head dip to deal with, and when he tried to stand straighter at impact, the club still passed his hands and he started hitting tops and chunks. He needed to dip down to get the club to the ball, and now what the camera had shown him had made his swing—and his results—even worse.

Between junior and senior year, Gina became his wife. In the middle of senior year, they welcomed baby boy Ross to the world. Caring for an infant made playing in a tournament feel like a piece of cake, so Leitz's golf game improved. But by the time he graduated with a marketing degree in 1981, he knew playing professional golf wasn't an option. With a budding family to provide for, he took a job for $4.25 an hour as an assistant superintendent at Sherwood Forest Country Club in Baton Rouge. Within two weeks, the job as head professional at Pinewood opened up, and for an annual salary of $10,800, Leitz jumped at the prospect of returning.

He showed up to the interview with all of his modern university skills on display, bringing a typed-up plan about how he would run the boys and girls junior program, manage the golf carts, and improve merchandise sales. The older men who had interviewed before him had shown up sun-beaten and ready to chat, and Leitz blew them out of the water. With interviewees still to go, the club offered Leitz the job. As a twenty-two-year-old with no prior teaching experience, Leitz began his life as the head pro.

Much to his chagrin, members started asking for lessons. Leitz tried as best he could, thinking he knew a lot, but always aware that the information he had obtained had never helped his own game one bit. Students would come and they would leave, no better than they were before. Leitz started feeling bad and spent a good part of his second year not at the club, but out on the PGA Tour, caddying for his former McNeese teammate Tim Graham, who tied for medalist at the 1981 PGA Tour qualifying school.

By 1982, Panasonic had come out with a portable video camera, and Leitz thought that would be the future of teaching. He spent $2,800 on it, and Gina was furious. He recorded himself and saw the same image of the clubhead in front of his hands. He recorded his students, and again there it was. No one was getting better, but he didn't know why.

Leitz thought it would be a good idea to invite Graham to Slidell so they could videotape his swing. Excited at the prospect of modern technology, Graham drove all the way from Memphis. They spent an afternoon filming, and Leitz would note which shots were pulls and pushes. When the replay started, his heart sank. He had no idea by looking at the video which shots were which, and he realized he had no idea what was going on. Of course, he didn't tell this to his friend, the PGA Tour player whose career and livelihood were on the line. Instead, Leitz buried his embarrassment and made some vague statements. Graham left impressed with the camera, pleased with seeing his swing from an outside perspective, yet still no better. By the end of the year, he would lose his Tour card.

After that, Leitz stopped giving lessons. He thought that if he was going to take people's money, he had to know what he was doing. He wasn't helping people get better, and he wasn't about to become a con man.

In 1986, he spent another small fortune on another video camera; this one could shoot sixty frames per second. He saw the same image of himself, and the same image of other amateurs on the range. He still didn't know what was wrong.

Then the PGA Tour came to Lakewood again, and Leitz decided to bring his camera to a practice round. Through connections as a head pro, he got himself inside the ropes, and he headed straight to the range to take head-on videos of Nicklaus and Lee Trevino, two of the best ball-strikers of all time. As Leitz stared at the grainy footage, life slowed to a halt. He saw both men make contact with their hands in front of the ball, and the shaft leaning toward the target. It was the opposite of the position James had, and it was all so simple.

*Son . . . of . . . a . . . bitch.*

His heart raced as he moved down the range with his camera, thinking, pacing. *It must only be the really good players who do this.* He taped a club pro he had never heard of, followed by an Argentinean named Vicente Fernández. There they were, hands in front of the ball, shaft leaning forward.

"That was the best day of my life," Leitz says, looking up from his black chair to that wall, covered with framed photos. There the pictures hung, the moments of impact immortalized—Nicklaus, Trevino, Hogan, all flushing irons; Tiger Woods, Fred Couples, Laura Davies, all booming drivers. All of them, hands in front of the ball.

It's a hanging testament to the truth that it took Leitz so long to learn, a truth that never became evident until he had worked hard enough to earn its discovery. He looks at his computer screen, with the photo of himself from college, and looks upon the wall with eyes wide open, his eyebrows raised in a sense of astonishment, of retrospective realization. This is no longer about the technical, or the mechanical, but about raw emotion. This look is the physical description of human longing, and perseverance, and eventual success. It's about a man in a shack in Louisiana believing deeply in the concrete world, and whose faith was mercifully rewarded.

On the wall, just to the left of the group of pictures, is another framed treasure. Once James had his moment of clarity, he began to dig even deeper into the world of scientific instruction, a dig that can only go so far before hitting bedrock. In that frame on the wall is a small business card with a strange green symbol that says James Leitz is a certified instructor of *The Golfing Machine*.

"The first time I read that book," he said, "I threw it against the wall."

# FIVE

# BABE RUTH AND TEACHING'S EVOLUTIONARY JUMP

As golf became a more quantifiable game, a strange word emerged into the lexicon: *own*. Tiger Woods famously said that only two players in the history of the game *owned* their swings—Ben Hogan and Moe Norman. But how can a person *own* a motion?

The connotation is of an ingrained and unbreakable confidence in mechanics. It is the reaching of the point where no outside influence—say, a coach—is needed. Yet no player has ever made two exactly identical golf swings, nor should anyone want to. The variables of the extremely complicated bodily motion and the variables of the external circumstances surrounding each shot are infinite. For one, the proprioception of each player varies from inexact to woefully inept. The best players and teachers have come to the hard truth that golf is intrinsically an inconsistent game.

Yet the idea of *owning* a physical motion remains, a reflection of a mind-set with a modern, scientific bend. If you can own something, then it is separate from you. It is detached, an object you can lose, if you ever had it at all. But as an outside object, you can also strive to obtain it, through study or hard work or some other avenue of purchase. This point of view was exemplified with *The Golfing Machine*, then went through a monumental jump forward in the mid-1990s, as shown in three consecutive Masters tournaments. The way the golf world perceived coaches drastically changed from Ben Crenshaw's heart-wrenching victory in 1995, to Nick Faldo's technical masterpiece and the unraveling of Greg Norman in 1996, to Tiger Woods's victory in 1997, the kicking-off point for the biggest shift in the sport's history. But where was the starting point? When did coaches become necessary characters?

Before Crenshaw's 1995 win, teachers were generally an afterthought. Just as so many before him, Crenshaw rarely thought of mechanics. He was the ultimate feel player, working with his childhood teacher, Harvey Penick, only sporadically once he left for college, just as Nicklaus did with Jack Grout. Yet Faldo worked tirelessly with David Leadbetter not just to turn his career around, but to sustain his technical mastery throughout his prime. Then, after Woods and his cycle of instructors colored the game with such a rainbow of information, coaches became integral parts of any professional's success. In that small window of three consecutive tournaments at Augusta National just before the turn of the twenty-first century, a new era of golf began.

In this historical context, the golf swing evolved from being thought of as a malleable part of each person's nature, to being envisioned as an entirely blank canvas. Golfers became programmable machines, and their hardware and their outside programmer were considered the largest determinants of their success.

Amazing how just one word entering the lexicon—*own*—can

speak volumes. Even more amazing how that mind-set began—with one of the best baseball players ever.

From 1929 to 1934, Babe Ruth had a backup on the Yankees named Sam Byrd. Often referred to as "Ruth's legs," Byrd would come in for the prodigious slugger late in games as a defensive replacement or a pinch runner.

Byrd carved out a nice little career, starting when he hit .312 in 203 plate appearances in his rookie season. Yet the native Georgian always loved golf, and in the middle of his career he won a tournament of all baseball players in Sarasota, Florida, by 14 shots, shooting even par over four rounds. Right after Bobby Jones won the Grand Slam in 1930, the Yankees came to Atlanta and Jones asked Yankees manager Joe McCarthy if Byrd could sneak out for a round of golf. According to a story the *Birmingham News* later published in 1937, Byrd used borrowed clubs on a course he had never seen and shot 1-under par, tying Jones. When asked what he thought of Byrd's game, Jones said, "He's the best man off a tee I ever saw." When asked if Byrd really was "one of the best," Jones clarified, "No, not one of the best; the very best man with a driver I ever saw."

Byrd traced his proficiency back to Ruth, who taught him how to hit a baseball when he was first called up in 1929. Ruth explained a drill he used where he kept a handkerchief under the lead arm—in Byrd's right-handed stance, that meant his left arm. The goal of the drill was to keep the swing of the bat level, avoiding an uppercut or a downward chop, which would both result in poor contact. Ruth added that all good hitters "brace" on their back leg, coil around it, then fire through with big muscles—meaning the core muscles between the hips and shoulders—transferring their weight to the front leg. It's exactly what a grainy video of Ruth hitting a home run looks like. By keeping the lead arm in close to the body, the whole thing stays "connected." The drill is still used across sports today, with Tour players often using headcovers instead of handkerchiefs.

Byrd used Ruth's methodology with some success in baseball, and when Byrd thought about the golf swing, all he did was change it to an inclined plane. The image was that of a flat table for the baseball swing, then tipping it over to about a forty-five-degree diagonal for golf so that the one edge faces your shoulders and the other faces the ball. To Byrd, that made the baseball swing and the golf swing identical, just on different planes. One day while Byrd was playing golf with famed newspaper columnist Grantland Rice, the idea for a collaboration on a book about the differences in the two swings was born.

"Well, Granny," Byrd said, "it's going to be a darn short book."

Byrd eventually quit baseball in 1936 after a bout with malaria and took on a job with Ed Dudley, who worked as the head pro at three-year-old Augusta National Golf Club during the winters and at Philadelphia Country Club in the summers. Despite battling the putting yips, Byrd still managed to win the 1939 Philadelphia Open (one year before it was named an official PGA Tour event), then finished third and fourth, respectively, in the 1941 and 1942 Masters. He won his first of six Tour victories at the 1942 Greensboro Open, and the highlight of his playing career came when he made the final of the 1945 PGA Championship (then a match-play event), losing to Byron Nelson in the midst of Nelson's eleven-tournament winning streak. Byrd was a popular guy on the Tour, and he used to make trips down to Florida with Ben Hogan to see their friend "Wild" Bill Mehlhorn.

By 1960, Byrd was giving lessons at a driving range and par-3 course he owned in Birmingham, Alabama. That year, he hired a gangly seventeen-year-old named Jimmy Ballard. Their mutual acquaintance was a local pro named Lee Mackey Jr., famous for his record 6-under 64 in the first round of the 1950 U.S. Open at Merion Golf Club, at that point the lowest single-round score in major championship history. (That was the same tournament Hogan won in his trium-phant return after a near-fatal car accident.) Ballard began taking les-sons with Byrd and working in the shop, once picking up the phone to take a message for Byrd.

"Please tell him to call Ben Hogan when he gets back," the voice said.

"Hell, I thought it was a joke," Ballard remembered. "So, if any-one ever asks if I spoke to Hogan, I can say I did—when I was sev-enteen!"

By 1966, Byrd had ceded control of the driving range to Ballard, who had begun teaching and saw his student load continue to in-crease as he kept using Byrd and Babe Ruth's idea of "connection." Ballard kept putting handkerchiefs in people's armpits and watching them play better. By 1970, Ballard was becoming incredibly popular in the local area, and a student of his referred a player named Mac McLendon. Actually, the student told the struggling McLendon that he would give him $50 if he just went and saw Ballard.

McLendon had played golf at LSU and came out in 1967 as a first-team All-America. He won the first pro tournament he played in, the 1968 Magnolia State Classic, a satellite event played opposite the Colonial Invitational. McLendon went to Memphis the next week, and he said, "I should've won that tournament, too." Playing in the final round with his boyhood idol, Arnold Palmer, and the great Lou Graham, McLendon finished third.

"I was off to a roaring start," he remembered. "But I knew, even though I had chances to win two or three more tournaments in the second half of 1968, that when I went back and assessed why I didn't win those tournaments, it was really that I didn't have the length off the tee."

McLendon wasn't sure how to go about hitting farther, and a half century later he laid out a fact that remains true: "You can get a lot of lessons on a practice tee on Tour. I just got myself all fouled up."

So he went to see Ballard, put the handkerchief in his left armpit, and immediately he started hitting the ball twenty to thirty yards farther. Even though his tournament results were inconsistent, the two kept plugging away, and eventually McLendon teamed with boyhood friend Hubert Green for the 1974 event at the Walt Disney Resort in Orlando. Holding a one-shot lead, Green birdied the seventeenth hole, and then they came to the eighteenth of the Magnolia course and McLendon hammered a drive through the fairway. He remembered playing this tournament in years past and needing to hit a 3- or 4-iron into the green. He was now left with a 9-iron, and he stuck it, winning his first big event on Tour with a combined total of 33 under.

"That is when it dawned on me," McLendon said. "I do have it."

That win made people take notice of McLendon, a good redemption story of a player who lost his game and then rediscovered it—and all the while McLendon was giving credit to Ballard. Players were intrigued by this teacher who could succeed with such a reclamation project, and even more so after McLendon racked up his first solo victory in 1976 and two more in 1978. Players started coming to Ballard en masse, and he eventually moved his operations to the famous Doral Resort in Miami. Now he wasn't getting fringe players trying to scratch out a living, but top-tier players who thought they were getting left behind without a big-name teacher as support. Ballard took on a cocksure eighteen-year-old named Hal Sutton, the best amateur in the country in 1976, who had only one goal—to beat Jack Nicklaus.

"Mac McLendon's mind-set when he came to me was that he wanted to play the Tour to see if he could win," Ballard said. "Now Hal Sutton, he came to me as an eighteen-year-old out of Louisiana and said, 'I want you to understand—the first thing I want to do is beat that guy from Ohio.'"

In contrast, Nicklaus himself would check in with his lifelong instructor, Jack Grout, about once a year before the season started. The two had worked together since Grout left Glen Garden in Fort Worth, Texas—the course where Hogan and Byron Nelson caddied as youngsters—and moved to Scioto Country Club in Nicklaus's native Columbus, Ohio, in 1950. At the time, little Jackie was ten years old, and as would be the case for the remainder of the best career in pro golf history, they worked strictly on the fundamentals, with no real overarching "method." Grout's only goal was to lay a foundation and let Jack carve out the rest according to his individuality.

"Jack and I both know countless promising golfers who have become hopelessly confused through failing to learn these fundamentals at the outset, usually with the result that they start confused and then compound the confusion by switching from method to method or teacher to teacher, until eventually they end up trying to play a dozen different ways all at once," Grout wrote in the foreword to Jack's instructional manifesto, *Golf My Way*. Many years later, the statement still rang true when looking at all the pros caught up in differing mechanics.

"Jack never fell into that trap," Grout wrote, "and I believe that his evasion of it is one of the less-recognized factors behind his greatness."

Even when Nicklaus was excelling at Ohio State, he would check in with Grout once in the winter, as the two would hit out of what Grout called "an open-ended Quonset hut."

"Jack Grout never set foot on the practice tee, never one time when I played golf," Nicklaus said. "Was Jack Grout there a lot? Yeah, he went to a lot of golf tournaments. Jack Grout was back in the bleachers, and if I wanted something, I just walked back in the bleachers and said, 'What do you see, Jack Grout?' He'd say, 'Your head position is a little off.' And that would be about it. It was a pretty simple thing."

It wasn't very different from Nicklaus's contemporaries, either. Arnold Palmer grew up the son of a greenkeeper nicknamed Deke in western Pennsylvania. Palmer's curlicue swing was wonderfully original, never groomed by any sophisticated instructor nor used as a teaching template after. Lee Trevino grew up the son of Mexican immigrants in Texas, and his short and choppy swing that produced such pure ball-striking was honed while playing for more money than was in his pocket while serving in the Marine Corps in Okinawa, Japan. Seve Ballesteros fed his peerless imagination by growing up with just a 3-iron, learning how to hit it in every situation off the sandy beaches in his native Spain. Tom Watson was the last player to challenge Nicklaus for supremacy in the game, and Watson's languid swing was initiated by Stan Thirsk at the Kansas City Country Club and changed little through Watson's long and illustrious career, as he was still competitive at the highest level into his sixties.

These were all headstrong men, hardly relying on anyone but themselves to prepare for competition. Not a lot of attention or thought was put into their mechanics. Each of their golf swings was originally their own, part of their personality. It was part of what made their competitions against one another so enthralling, that they were such different people who played the game in such different ways.

While not immune to the desire for perfection, they didn't think that teachers had the answers. They all grew up with Hogan as the cultural sporting idol, with his mystique of technical mastery and secrets unlocked in hard work. Living up to the iconic Hogan template was an ideal that Nicklaus, for one, realized was impossible—a realization that saved his career before it had even started.

"As an amateur, there were times when I believed that if only I didn't have to clean up my room, or get an education, or earn a living, I would be able to hone my game to point of absolute perfection and then hold it there permanently," Nicklaus wrote. "I grew up in the era of Hogan. Everything I saw of him and read of him and heard of him indicated that he had achieved utter mechanical perfection in the striking of a golf ball. Perfect repetition. Flawless automation. This was my dream. All I needed to achieve it was sufficient time to work at my game.

"I was kidding myself. When I turned professional, suddenly I had all the time and opportunity I needed. And I discovered, fast, that my dream was just that: a dream. No matter how much work I did, one week I would have it and the next I couldn't hit my hat."

Acceptance of that inconsistency is also part of what made Nicklaus so great. The inability to accept it, or the driving desire to eliminate something that was inherent in the game, was what drove so many people so deep into instruction. By way of McLendon's success, Ballard was the teacher whom everyone wanted to work with. His method had an interesting origin in Babe Ruth, and anytime pro golfers could consider what they did more athletic and more in line with other sports, the more they liked it. Ballard continued to help players reach success, and in 1988 he achieved what he liked to call the "Teachers' Grand Slam." At the time, Ballard was teaching Sandy Lyle, who won the Masters; Curtis Strange, who won the first of his back-to-back U.S. Opens; Ballesteros, who won the British Open; and a Swedish kid named Christian Harden, who won the British Amateur—not quite the professional sweep, as the PGA Championship was won by Jeff Sluman, but surely an impressive bit of résumé building for Ballard.

He was named Teacher of the Decade by *Golf Magazine*, a newly coined title that would have been almost unfathomable in any decade prior. The celebration of the coach signaled a drastic shift in the nature of the teacher-student relationship. Ballard was so confident in his position that he got into a tiff with the PGA of America, the organizing body of all teaching professionals. He never became an accredited teacher because he never attended the new instructional schools, which he found to be a waste of time, if not entirely misleading in their prescribed methodologies.

"I'm teaching Johnny Miller, Gary Player—I'm busy as hell," Ballard said. "You want me to go to school?"

McLendon eventually retired from pro golf to become a

stockbroker—something he always dreamed of doing. Ballard continued to think he never got enough credit for the breakthrough work he
did. But the biggest impact was in the drastic change in how teachers
were viewed. Just one player got Ballard on the map, and the landscape of the game had started to shift.

The tremors of that shift were being felt as the 1995 Masters began,
a tournament that would go on to be defined by something distinctly
human and at least partially spiritual.

At the time, Ben Crenshaw's career seemed to be coming to a
premature end. The five feet nine, 160-pound wunderkind was a
three-time NCAA individual champion at the University of Texas, and
between turning pro in 1973 and 1995, he had won eighteen times on
the Tour, including one major, the 1984 Masters. Now at forty-three
years old, Crenshaw's game had fallen off. He hadn't won in almost a
year. He had missed three of his past four cuts on the Tour and hadn't
broken 70 in two months. The best putter on the planet—and arguably the best of all time—was ranked 69th on the Tour in putts per
round.

Crenshaw was also playing with a heavy heart. Two weeks prior,
he had been back in a small bedroom in Austin, Texas, where his
teacher and mentor, Harvey Penick, lay dying. Penick was ninety
years old, and for one last piece of advice, he had told Crenshaw to
go to the garage and get an old Gene Sarazen putter. Penick checked
Crenshaw's grip one last time, then told him, "Just trust yourself." A
week later, Penick died.

Penick started at the Austin Country Club as a caddie when he
was eight and was named head pro in 1923, the year Bobby Jones
won his first major. Penick put a club in Crenshaw's hands when he
was six, just as he did to Tom Kite soon thereafter. Penick's teaching
style was grandfatherly. He told stories and created analogies to show

his students how to execute rather than telling them. When Crenshaw and Kite were budding junior players, they asked Penick how to hit a high pitch shot over a bunker to a tightly tucked hole. Penick sent them to the practice green, where he told them to pretend that a large tree was growing out of the sand. That tree should grow, Penick said, until it was high enough that the ball landed softly on the green and stopped. Penick left. When the two young boys ran into the pro shop to say they did it, Penick came back out. He watched, and they asked him the technique for how to do it. He told them to do it again, then said, "It's what you just did." There wasn't a single mention of mechanics, and the lesson stuck.

"An old pro told me that originality does not consist of saying what has never been said before" is how Penick led the introduction to his cornerstone of golf literature, *Harvey Penick's Little Red Book.* "It consists of saying what you have to say that you know to be the truth."

Penick's lasting legacy is partially contained in this tight little collection of anecdotes and teaching notes that he took throughout his career. For a long time, he only kept it for himself, with no plan to ever share it. But as he grew older, his students convinced him it was only right for the world to glean his insights. In 1992, when Penick was eighty-seven years old, the *Little Red Book* was published, and it became a huge bestseller. He described his teaching method as employing "images, parables and metaphors that plant in the mind the seeds of shotmaking." He referred to Kite as "Tommy" and said that Crenshaw was such "a natural" as a teenager that Penick's main job as a teacher was to keep him from practicing too much and messing himself up.

The original edition had numerous introductions from students, including one from Crenshaw: "For all of his admirable traits, let us simply say that Harvey Penick presents the very best that life and golf can offer." The one from Kite read, "The one thing that we all have learned from Harvey is love. A love of a game that teaches us more about ourselves than we sometimes care to know."

With Penick's funeral set for Wednesday of Masters week, Crenshaw was on the range on Tuesday just flailing away at balls, his malaise growing deeper. His longtime Augusta caddie, a lumbering man named Carl Jackson, felt compelled to tell Crenshaw to put the ball back in his stance and turn his shoulders just a little bit more. Crenshaw hit four more balls and knew he had just found the answer.

"I've never had a confidence transformation like that in my life," Crenshaw said, according to the masterful *Sports Illustrated* game story written by Rick Reilly.

At 7:30 a.m. on Wednesday morning, Crenshaw and Kite flew to Austin for the funeral, where Reilly wrote they "carried a very light box and their own heavy hearts to the grave." Penick's son Tinsley gave the Sarazen putter to Crenshaw, then he and Kite flew back to Augusta.

Kite missed the cut, and the tournament that Crenshaw played was hardly flawless. But he seemed to be getting such a flood of good breaks—starting with Jackson's advice—that it was logically startling. In the final round, he hit a wayward drive on the par-5 second hole, and it bounced off a tree and right into the middle of the fairway. *Harvey.* After rolling in a tidy little birdie at thirteen, Crenshaw was under a tree on fourteen and hit a punch 8-iron that bounced off a mound and settled twelve feet away. *Harvey.* Davis Love III, tied for the lead on sixteen, hit a 7-iron that went five yards farther than he could have thought possible, and it stayed up on the top tier of the green and resulted in an almost-unavoidable three-putt. "Sometimes you wonder if things are meant to be," said Love, who had his own connection to Penick and wanted to attend the funeral, yet was told to stay at Augusta and practice by none other than Crenshaw.

Greg Norman, the best player in the world at the time, had his chance ruined by a 106-yard sand wedge into seventeen. He pulled it forty feet left and three-putted, taking him out of contention. Love birdied seventeen and finished with a splendid 66, putting him in the clubhouse at 13 under, tied with Crenshaw as he came to the par-3 sixteenth. There, Crenshaw hit it to five feet and knocked the putt in for a birdie. He then made a 12-footer for birdie on seventeen and went to eighteen with a two-stroke lead.

He made an emotional bogey on the closing hole and won by a shot, dropping his putter, folding his head into his hands, crying and burying himself into the hulking embrace of Jackson. It was as tear-jerking a win as golf has ever seen.

"It was kind of like I felt this hand on my shoulder," Crenshaw said, "guiding me along."

The hand guiding Nick Faldo in 1996 was hardly a ghost.

David Leadbetter was the next in the line of Ballard, taking the role of celebrity teacher to staggering heights. Like Ballard, Leadbetter inserted himself into the conversation and became an important piece of the landscape wherever his students were playing. Yet Leadbetter's attractiveness was hardly because of a singular methodology, but rather his extremely analytical approach to teaching mechanics.

His early ideas were about breaking down the golf swing into positions. Leadbetter had read and loved *The Golfing Machine*, but he only paraphrased the physics. He also boiled down the countless variations and found that having the club and the body in certain places throughout the swing would produce the most consistent results. He liked the spine at a certain angle at address, then the hands and the club coming back directly on plane, with the clubface first perpendicular and then parallel to the plane; same for the top of the swing, when he wanted the clubface and left arm to parallel to the plane. The transition from backswing to downswing was achieved in another sequence of positions on the way down, and from the view behind the player—"down the line," as it's known—the clubhead was released ("turned over") and reappeared on the other side of the player's body, on plane again. Leadbetter had become so popular that by the 1996 Masters, these specific mechanical positions were considered the standard. By being so detailed, and with video emerging as the tool to compare and contrast with such precision, Leadbetter had created a mechanical orthodoxy, likely the largest consolidation in the history of golf instruction.

Leadbetter was tall and angular and often wore a signature wide-brimmed straw hat. His family left Worthing, England, when he was young, and he grew up playing all sorts of sports in South Africa. When he was fourteen, he went to an exhibition in Rhodesia—later known as Zimbabwe—and shagged balls for South Africa's favorite golfing son, Gary Player. He attended PGA instructional seminars in America in the 1970s and competed in some South African PGA events, rarely making a cut. By 1977, his family had moved back to England, so he decided to give a go at the European Tour qualifying tournament. It was November, cold and rainy at Foxhills Golf Club, about ten miles south of Heathrow Airport outside London. By the back nine of the final round, he knew he wasn't playing well enough to be in contention. His mind relaxed, and somewhere around the twelfth hole or so, he dunked his second shot on a par 4 for an eagle. He birdied the next, made another birdie along the way, and had a 10-foot putt on eighteen for what would have allowed him to qualify right on the number.

"It was like, if you've ever had an easy putt, this was an easy putt," said Leadbetter almost forty years later. "Up the hill, right to left—and damn if I didn't leave this thing right on the lip. Literally, one more roll, this thing was in. My God.

"It's funny how things work out. Probably for the best, anyway."

Leadbetter became a club pro in England and kept in touch with a childhood friend from South Africa, Nick Price, teaching him informally and sporadically. The two had competed against each other as juniors, but Price began to excel and was making a living on the European Tour. Both wanted to move to the United States for the better weather and better opportunity, and eventually Leadbetter ended up at Grenelefe, at the time a premier resort just outside Orlando. By 1981, Leadbetter's video camera had become ubiquitous, and Price started working with him full-time. By 1982, Price held a three-shot lead with five holes to play in the Open Championship—the same Open Championship, eventually won by Tom Watson, that Bobby Clampett blew in the haze of mechanics. Leadbetter's persona grew, and in 1984 he was down in South Africa with another childhood friend and student, Denis Watson. There, he ran into Faldo, who was twenty-seven years old and a ten-time winner on the European Tour. He had just won his first PGA Tour event at Hilton Head Island, but he wanted more out of his career, so he asked Leadbetter to take a look.

"You've got about half a dozen things wrong with your swing that I can spot," David said, according to Faldo's candid 2004 autobiography *Life Swings*. "But rather than trying to tackle each one individually, I'd suggest we try to find the one thing that will end up curing them all in a chain reaction."

"His words made immediate sense," Faldo wrote.

At Nicklaus's host tournament in Ohio the next summer, Faldo walked up to Leadbetter on the range and said, "I'm all yours. Throw the book at me."

Faldo said he wanted "to be number one and win an Open" and had "no timetable" for how long it would take to be that good. Leadbetter responded by saying Faldo's swing was "old style" and "too willowy," and that it would take two years to rebuild it. Both men were analytical and loved to cram themselves into a little room and watch the tape of Faldo's swing, drawing lines and circles on the screen with markers. "A picture is worth a thousand words, right?" Leadbetter said.

Yet that was all background information for Faldo. He wanted it all to be mechanically definitive so he could trust it, but when it came to implementing the new moves, he knew he couldn't compete if conscious thoughts were flooding his mind. If he and Leadbetter were working on keeping the flex in Faldo's right knee, or pulling back his left shoulder, Faldo would put them into a little singsong.

"Sit and pull," he'd say to himself. "*Sitandpull, sitandpull, sitand-pull.*"

It was always two or four thoughts—never three, never one, never five. Maybe the numbers were about rhythm, but that was never discussed. All that was clear was that if Faldo was playing well, it was two. *Sit and pull.* If he was struggling a bit, it was four. *Sit, pull, coil, and release.* But during the prime of his career, it was never anything different from two or four. Sometimes, Faldo didn't even use words. It might be sound—say, *Swooosh* of the right hip, or the *Thump!* of flushed iron—fitted into the singsong of the moment. *Sit and swoosh. Pull and thump.* It was a cadence. It was an athlete who intrinsically understood what worked for him and what didn't and knew how to transform theories into an action. Faldo always thought of himself as a feel player and never once thought he focused too much on mechanics.

"He was actually sort of a genius," Leadbetter said, "at being able to incorporate swing mechanics into feel."

The day after his thirtieth birthday, just over two years since he started working with Leadbetter, Faldo made 18 straight pars in the final round of the 1987 British Open at Muirfield in Scotland and won. It was the defining moment of his career, the first of six majors. More than anything else, observers were struck by his cool (if not cold) demeanor, and his exacting precision that came via his relationship with this rising teacher. The view was that Leadbetter had torn down a good player and rebuilt him as a great one, just as Ballard had done with McLendon.

It was more than just a trend, and every player who was down on their game—which is every player at some point—went looking for a new teacher, one not just with a different perspective, but one with the definitive answers such as Leadbetter and his video camera. Some of the power in the game had shifted to the people with more tangible information, and those were the instructors.

It was the same thing for Greg Norman, nicknamed the Great White Shark, the tall, white-blond Australian with the straw hat, who might have been as physically gifted as anyone to play the game. Yet by 1991, Norman's promising career had been defined by a series of heartbreaking defeats, and his game was going sideways after he missed the cut at the Masters for two straight years. In a nonsanctioned, big-money event in Jamaica in December of 1991, Faldo drained a 15-footer on the seventy-second hole, then Norman missed a 4-footer for a birdie that would have won the tournament. Norman had still shot a final-round 63, without a bogey, and yet on the first hole of the playoff, he found a greenside bunker, couldn't get up and down, and remained winless for almost two years—it being made even more poignant by Faldo being the one who beat him.

Norman walked through the heavy Caribbean air back to his hotel room and stared at himself in the mirror. Having been a disciple of Zen teachings and motivational speakers such as Tony Robbins, he began to ask himself questions. *What do you want to do? Do you want to give up the game?* The thirty-six-year-old was forcing himself to contemplate these honest questions. His answers were that he wanted to fight and keep competing. He didn't think he was done. He thought he had more to give.

So he did what Faldo and McLendon had done before him and found an instructor to help rebuild his swing.

At the time, Butch Harmon was working at the Lochinvar Country Club outside Houston, and he was teaching Norman's fellow Australian Steve Elkington. Norman had a long-standing friendship with

Elkington, who was known as whip-smart, opinionated, one of the most elegant swingers of the club, and a pure ball-striker. Elkington left Australia to come play for the powerhouse team at University of Houston and stayed in and around the area for most of his life. He would get immersed in swing theory, especially *The Golfing Machine*, and was often seen practicing just with Ben Doyle's Facts and Illusions mat. Later in life, after his career peaked at the 1995 PGA Championship when he beat Colin Montgomerie in a playoff at Riviera, Elkington would take lessons from just about every teacher anyone has ever heard of. In 1991, Elkington saw his buddy Norman struggling and said that it might make sense if he got together with Harmon. The two had similar laid-back and friendly personalities. At the Houston Open that year, Norman approached Harmon and asked him to take a look at his swing.

"Sure, Greg," Butch said. "You're one of the best 1-iron players in the world. Take a 1-iron and hit me a high fade."

Norman did.

"So what's going wrong?" Harmon asked with a smile.

"Jesus Christ" was Norman's candid response. "I like that."

That set the tone for their relationship. Less technical a teacher than Leadbetter, Harmon was there as a guide, a signpost. That was exactly what Norman was looking for after the way he lost in Jamaica, and that long and literal look in the mirror.

"First of all, I never wanted a teacher," Norman said decades later. "For me, a teacher is somebody who wants to teach you their way, their method. A coach is a totally different deal. A great coach will work with you on everything you've got going on. Your physical ability, your physical ailments, your flexibility, your life, understanding your personality, how to inject their personality, how to give you confidence when you're feeling down—that's a great coach. And there are very few great coaches out there."

By late 1992, Norman had yet to win again but was feeling a

little better about his game as he and Harmon headed up to Canadian Open at Glen Abbey. They had a plan for that week—or, as Greg called it, "a mission."

"That mission was just to get back to our old ways, whether that was physically or mentally, just get back to our ways. We worked together on a plan on how to conquer this . . . this *shitty* mountain we were on."

With the help of Harmon—and with a young Sean Foley looking on—Norman stole the show that week. After birdieing the seventy-second hole to get into a playoff with Bruce Lietzke, Norman hit a 240-yard 3-wood into the back bunker on the first playoff hole, then got it up and down for his first victory in twenty-seven months.

"It was a big step for me," Norman said. "A huge step. I had almost forgotten what it was like to win."

Norman went on a roll from there, winning his second Open Championship in 1993. He dominated the sport for the next two years, and in August of 1995, he made a 70-foot chip at the World Series of Golf to beat Billy Mayfair and Nick Price in a playoff, overtaking Price as the No. 1–ranked golfer in the world, a position Norman would go on to hold for a then record 331 weeks.

Norman was the hottest player in the world going into the 1996 Masters, Augusta National being a place where he had suffered his most brutal heartbreaks.

"This is the greatest championship around," Norman said in his pretournament press conference. "There's no other golf tournament anywhere in the world that generates the type of feeling like here at Augusta National. Any golfer, no matter what his stats or position in the world, whether a budding amateur or a professional, we all want to win it."

Yet it was impossible to think of Norman without thinking of all the gut-wrenching losses he had suffered over the years. At the 1986 PGA Championship, Bob Tway holed a bunker shot on the seventy-second hole to beat him. At the 1987 Masters, a local kid named Larry Mize holed a 120-foot chip shot on the second playoff hole to win. "That one probably stung me the most," Norman said, "because at no stage in my mind did I ever think I was going to lose that tournament. At no stage. And when that shot when in, I went, '*Ugh!*' And this is when I started telling myself, 'Always expect the unexpected.' So, that's a motto that's been with me every day since."

Norman bogeyed the seventy-second hole of the 1989 Masters to miss a playoff between Faldo and Scott Hoch, and later that year he lost to Mark Calcavecchia in a playoff at the Open Championship. Appropriately, Norman became the first player in history to lose in a playoff at all four majors.

"I do believe certain things are meant to happen for you," Norman said before the 1996 Masters. "Sometimes, you don't feel like you have a chance to win and, boom, something happens. You get your good breaks and your bad breaks. But I like to feel that things are meant for a reason."

Other heartbreaks fit the script, as well, even if they were on lesser stages. Journeyman Robert Gamez holed a 7-iron from 176 yards over water on the seventy-second hole to beat Norman in the Nestlé Invitational in Florida in 1990, and a few weeks later, David Frost holed out a bunker shot on the last hole to beat him in Louisiana. It all just seemed part of the tragic persona that Norman had acquired.

"God never gives a golfer everything; he always holds something back," Lee Trevino once famously said. "Jack Nicklaus didn't get a sand wedge, and Greg Norman didn't get any luck."

But still, Norman went into that 1996 Masters with a head full of confidence, and it showed. He put on a masterful and artistic display of driving off the tee, the first player to overpower Augusta National. Yet Norman had been dealing with some lingering back pain since the previous Wednesday, when he went to the practice tee and couldn't take the club back, having to cancel his practice round. He was working with his physical therapist, Pete Draovitch, and got it loosened up for Thursday's opening round, when he made nine birdies and tied a course record with a 9-under 63. He backed it up with a 69 and a 71, getting to 13 under, six ahead of Faldo going into the final round.

But after doing hours of his media obligations Saturday night, Norman woke up on Sunday with more tightness and aches. As was often the case, the SI joint in his hip had gotten locked out of place, creating stiffness in his back that couldn't be massaged out. After hitting so many balls that week, walking up and down the hills on the soft grounds of Augusta, Norman was struggling badly. Draovitch couldn't get it out, and a mile-and-a-half walk that Sunday morning didn't help either. Norman knew he was in trouble when he met Harmon at the range.

"It's not going to be an easy day," Norman said. "My club is going to be stuck behind me because I can't free up my hip action."

He started hitting balls, and Harmon did all he could to keep his prized pupil focused and feeling good.

"Everything looks great," Harmon said. "It's not stuck . . . it's not stuck."

"Butch," Norman snipped back, "don't bullshit me."

"He was definitely a different person, physically and emotionally," Harmon would tell *Golfweek*. "He fought his back all week, but played within himself. Sunday, it was like he tried to push everything. There was a tremendous amount of anxiety in his body that day."

Norman was trying to push aside the anxiety and the physical pain, but he would always disregard that as a reason for what was to come. Even after that conversation with Harmon, he laughed with fellow competitor Frank Nobilo, taking time from hitting balls to set up a date for a future practice round. Norman poked a finger into Harmon's head, drawing a big laugh from the two of them and caddie Tony Navarro. Norman tried on three different shark-emblazoned gloves before finding one that felt just right.

"I remember waking up feeling hopeful and relaxed," Norman even wrote in his 2006 biography, *The Way of the Shark*. It was a poorly veiled, if admirable, attempt to avoid making an excuse. Once Norman and Faldo got to the first tee for their 2:49 p.m. start time, the next couple hours were a study in tragedy. Norman made a bogey at the first, and on the tee of the second, Faldo said, "I could feel the nervousness emanating from Greg. He gripped and regripped his club time and again, as though he could not steel himself to hit the ball."

Norman knew he was hurting, and after bogeys at the fourth and the ninth, the lead over Faldo had gone from six shots to two at the turn. The thin line between success and failure had never been so clear.

"If you're three or four feet out [of position] at Augusta National, you're going to pay the price," Norman said. "I was that way on the front nine. I hit what I thought were pretty quality shots, but they came up three, four, five feet short of the target. Like the one on nine, for example. I was three feet from having a tap-in birdie, and instead it [spins] and rolls down the hill and I walk off with a bogey.

"Your body knows. You can't fight your body, and you can't fool your body. Just like you can't lie to yourself in the mirror, looking at yourself eye to eye."

Norman's bogeys at ten and eleven got Faldo into a tie, and then Norman double-bogeyed the devilish par-3 twelfth, dropping his tee shot in Rae's Creek, giving Faldo a two-shot lead. A birdie at the par-5 thirteenth kept Norman in contention, and the eagle chip he had on the par-5 fifteenth remains the iconic picture of the tournament, just lipping out and bringing Norman to his knees.

But a hook off the par-3 sixteenth into the water, and that was it. Mercifully, things ended with Faldo shooting a superb 67, one entirely overshadowed by Norman's ugly 78, one of the worst collapses in major championship history. On the eighteenth green, after it was all over, Faldo went over and gave Norman a big hug, something that caught many off guard. The two hadn't said a word to each other all day. That was standard for Faldo, who had done it many times before and was preceded by his reputation as a robotic golfing machine with little emotion or feeling. That turned out to be only part of his complicated makeup. "I don't know what to say," Faldo told Norman, tears in both their eyes. "Don't let the bastards get you down over this."

Norman then signed his scorecard and nobly trekked right into the press center, facing "those bastards" and beginning his long and painful conference with:

"Well, I played like shit."


Leadbetter remembered the first time he met Tiger Woods, giving him a lesson in 1992 at Bay Hill outside Orlando. "I remember seeing this kid—he was like an elastic band, like Gumby," Leadbetter said. "And he moved so well, so much speed."

Then Leadbetter remembered how his prized pupil, Faldo, reacted after Woods's victory at Augusta in 1997. Faldo seemed to know immediately when draping the green jacket over Woods's wide shoulders that it was the last time he would be defending champion. "When [Woods] won the Masters in '97, I think that was really the start of Nick's decline," Leadbetter said, "because he looked at this kid and said, 'My God, he's hitting it fifty yards by me.'"

Norman had his own early encounter with Woods when the latter was fifteen years old. At the behest of the marketing company IMG, the two played a casual nine holes together down at Norman's club in Florida. "I was impressed, to say the least," Norman said of Woods.

Woods was at Augusta as an amateur in 1996 when all of the Faldo-Norman drama was unfolding. He was supposed to play a practice round on Monday that year with Arnold Palmer, but Arnold canceled. Woods played with Norman instead, the two sharing a coach in Harmon. Norman, who was rightly known as the best driver of the ball in the modern era, marveled at Woods's length.

"I think he's longer than John Daly," Norman said at the time. "He flights the ball so well."

The two students of Harmon's played another practice round together on Tuesday, joined by former Masters champions Raymond Floyd and Fred Couples, and Woods was taken aback at how free they were with their advice.

"Those guys know the course like the back of their hand," Woods said.

Palmer had canceled his original practice round with Woods because he had something special planned for Wednesday. He invited Woods to play a round with him and Jack Nicklaus, ten green jackets between them. The three went around and had a joyous time, and the praise could not have been more effusive for the twenty-year-old Woods.

"This kid is the most fundamentally sound golfer that I've ever seen at any age," Nicklaus said. "Hits the ball nine million miles without a swing that looks like he's trying to do that. And he's a nice kid. . . . Arnold and I both agree that you could take his Masters and my Masters and add them together, and this kid should win more than that."

Woods played the first two rounds with the defending champion, Crenshaw. It was the last time over the next twenty years that he would play in the Masters and miss the cut.

Woods turned pro in late 1996 and showed up to Augusta in 1997 ready to play in his first Masters as a professional. Faldo played with him in the opening two rounds—Woods was still the defending U.S. Amateur champion—and watched as Woods started the tournament with a front-nine 40. He then came home in 30 and went on to finish the four rounds at a record 18-under par, winning his first major by twelve shots.

The color of Tiger's skin wasn't lost on anyone, the club having allowed their first black member just six years prior. Augusta's co-founder Clifford Roberts once infamously said, "As long as I'm alive, golfers will be white and caddies will be black." Roberts was eighty-three when he shot himself on the banks of Ike's Pond in 1977.

When Woods was on his way to Butler Cabin, where Faldo would again deliver that first green jacket, Tiger stopped his security entourage when he saw Lee Elder out of the corner of his eye. Elder became the first black man to play in the Masters in 1975, the year Tiger was born. "Thank you for making this possible," he whispered in Elder's ear as they shared a teary-eyed embrace.

The next year, Harmon would help Woods rebuild his golf swing. After one of the most important victories in the game's history, which created an entire new world of promise, Woods turned to a teacher to change and to improve. He would lead the conversation about golf for the next twenty years, and it seemed that all of a sudden Woods made mechanics interesting and important. As science progressed and information became more abundant and varied, Woods chased after more concrete answers. Using an advanced understanding of the physical world to avoid some of the old frustrations was the modern path to improvement. The game had taken a big step forward in its evolution, but underneath was the same competitive ethos that had always existed.

"I think I underachieved all my life," Nicklaus said. "I think that's why I got better. And I think that if you feel you're overachieving or getting more out of what you should get, then you stop working. I always feel like I'm never getting what I should be getting out of what I'm doing. So you've got to work harder to make sure you do that.

"What I mean by underachieve, I think everybody underachieves. I think you always think you can be better. And if you don't think you can be better, then your head is too large. You're scratching your ears on the wrong side."

If Hogan led Nicklaus to think that there was a mystical place of mechanical perfection, then Woods led the generation following him to believe that some quantifiable answers are necessary for improvement. Those scientific answers do not come from within, but can be found by tapping into the foreign knowledge of teachers. Woods continually searched outside himself, always looking for a new way to understand his own nature on a deeper level. He wanted to *own* his swing, and once that started to rust, he wanted to buy a new one.

The entire concept of how to improve changed. No longer could you advance by just working harder. You needed to work smarter.

# SIX

# TECHNOLOGY FOR PROFIT

We'll call her Sophia.

Because of archaic and murky NCAA regulations (as if there were any other kind) and for the sake of her privacy, it's best not to use the real name of this nineteen-year-old college golfer from Spain. She was a high-ranking amateur playing under scholarship at a major Division I university, having just started her sophomore year.

This was not her first trip to the Titleist Performance Institute, a sprawling thirty-three-acre tract in Oceanside, California, colloquially known as TPI. The acronym had come to stand not only for a physical place, but also for an idea, one that traveled in namesake conventions, that certified its disciples, and that carried an underlying corporate ethos. It implied cutting-edge science, from ball-flight monitors and 3-D motion analysis to the best golf-specific physical fitness, and a thorough mental and psychological evaluation.

At TPI, promising players such as Sophia could tap into the source of what had become the leading science-based game-improvement brand. TPI was where many proprietary methods were born and where new theories would be hatched. If a player wanted to get better by grabbing hold of something concrete, TPI was the No. 1 place in the world to go.

Sophia's day began in the corner of a massive gym, with weight machines, treadmills, ellipticals, and all sorts of other workout equipment strewn about. Mirrors lined most of the walls above a firm-cushioned floor. She was seated in a reclined office chair, electrical nodes the size of quarters stuck to her forehead with gel, and a little bonnet lined with more nodes placed snugly over her straight auburn hair. Wires coming out of all the nodes were connected to a machine that recorded her brain-wave activity. A woman holding a clipboard told Sophia to stare at a piece of white Styrofoam taped to a beige wall about fifteen feet in front of her.

The woman was part of a company called Neurotopia (later renamed SenseLabs), which received most of its work from the U.S. military. She asked Sophia to start by staring at the piece of Styrofoam for three hundred seconds without blinking. "If you do have to blink," the woman said softly, "try to blink a lot at once." That would make it easier for a section of the brain-wave results to be spotted as an outlier and tossed out while trying to establish a baseline reading for Sophia's neuro-functions.

After Sophia stared for three hundred seconds, the woman with the clipboard handed her what looked like an oversize controller for a video game, about six inches wide with a button for the thumb of each hand. In the middle of the controller were nine red lights, aligned in three rows of three. When the eight bordering lights were lit and the one in the middle was dark, making it look like a bull's-eye, that was Sophia's cue to push either of the buttons as fast as possible. When any other assortment of lights came on, she wasn't supposed to push anything.

The test lasted almost thirty minutes. At times, the lights blinked rapidly, every second or so. Other times, empty pauses lasted up to thirty seconds. It was draining to watch. The computer recorded all of the results and the reaction times from the test, and all of the readings coming from the nodes glued to Sophia's head. The readings and results would be matched to find out how her brain was reacting to different stressful situations. Her scores would then be compared to Neurotopia's database of tests, and a profile of Sophia's neuro-functions would be drawn.

"How did she do with neuro-strength?" the woman with the clipboard whispered. "How did she do with reaction speed? How did she do with stress recovery? How did she do with focus endurance? All of those things we want to know. It'll put out numbers, and then basically we'll see where she's better in some places, worse in some places."

After the test, Sophia was spent. "I'm tired, like, whew!" she said, smiling and giggling like the teenager she was rather than just a faceless plus-2 handicap. "At the end, I couldn't concentrate."

Seventy-two hours later, after Neurotopia erased all the anomalies, such as blinking, Sophia's results came back. The first to see the results was Sophia's coach, Lance, the human equivalent in energy to a double espresso, with the same strength of conviction. How Lance delivered all this information to Sophia—if he delivered it to her at all—was the tricky part. The psyche of a competitive golfer is fragile.

But Lance's job was comprehensive. When Sophia first came to TPI, Lance had asked her about her goals and aspirations; about her strengths and weaknesses, both in her game and mentality; about what she wanted to get out of instruction; and if she craved large swaths of information or ambiguous ideas about feel.

"There are two types of players," Byron Nelson famously once said. "Those that need to know a little and those that need to know it all."

The test showed that Sophia didn't score the highest in tests of mental fortitude, failing more often than the average person after an incorrect answer. "Some people refocus after getting one wrong, and some people get worse," the woman with the clipboard said.

Sophia had previously told Lance that she felt that her self-worth was linked to her golf game. Many competitive golfers feel this, and the pressure is especially high for an international student on scholarship. Lance was trying to get Sophia away from the idea that a bad shot meant she was a bad golfer or, worse, a bad person. He used cognitive techniques from a company called Vision54, started by two renowned teachers/mental coaches, Lynn Marriott and Pia Nilsson. One exercise had Sophia start a round with ten tees in her left pocket, then she would evaluate each shot during the round on commitment, not outcome. If she thought she committed just seven out of ten on the first drive, she moved three tees to the right pocket. If the second shot was nine out of ten, she moved two tees back to the left. The goal was to keep all ten tees in the left pocket, thus focusing on the one thing you can control—commitment, not outcome.

Lance decided that Sophia would not benefit from knowing the actual workings of her brain. If she knew that her brain was especially liable to collapse under stress, her awareness of such a mental defect could become a self-fulfilling prophecy. Two mistakes in a row on the golf course could snowball into a downward mental spiral. Such a crisis of confidence could have superseded any game with tees.

If Lance thought that Sophia craved the information and would grow from knowing the shortcomings of her physical brain, he might have talked her through it. They might even have brought in a TPI-certified mental coach to figure out a strategy for her to buckle down after a faulty swing and reset focus. Lance had done that with other students and watched them get better from it. Targeting athletes' mental skills was like targeting their physical techniques: identifying a weakness in a player's game—this time in brain mechanics, not swing mechanics—and working on it to get better.

Lance sat Sophia in his windowed office and talked with her about her game, about how she wanted to get better at scrambling and how she needed precision with her irons. Then he said, "Let's talk about your mind. What's going on mentally?"

"I didn't really do anything," Sophia said, speaking of her mentality as just another aspect of the game that she could train. She had been taught that the mental side of the game can be worked on in the same manner as swing mechanics and physical fitness. "I was just at home, practicing. I think what I needed was [more than] one good tournament, getting confidence. Because I totally lost my confidence last semester. You knew it. I was talking to you, and like . . . I knew I was a good player and not showing it."

"Correct," Lance said.

"I didn't do anything, really, it was just liking myself more than I used to before. Before, I was like, 'Oh, you're bad because the way you're playing.' And this semester, if I miss a shot, it's just like, 'Oh, it's golf.'"

"Wow. And I say *wow* because most girls and guys don't recognize that until their junior or senior year."

"When I'm under pressure, I feel like I can give more of myself. I can concentrate more."

"So you like pressure?"

"I like pressure, but not when I'm worried. That's what I told you last semester. I play good when I'm under pressure, but not when I'm under worry. If I worry about not doing it well, that's when I play bad. But when I'm like, 'Oh, I feel like this tournament is important—let's do it.'"

"Great."

Sophia spoke a little about her schoolwork, and Lance suggested it could be beneficial for her to have more of a social life and not focus too much on golf. He relayed a story of a LPGA Tour player who had benefited from that same advice. He also added the example of Zach Johnson, the diminutive Iowan, who, with the help of a couple TPI-certified professionals, had built a team of eight people around him. By enlisting the expert help of various specialists, from a swing coach to a financial adviser to a spiritual adviser, Johnson was able to win the 2007 Masters and launch his professional career forward.

Johnson's team of specialists exemplified the organized, cerebral approach to drawing the most out of each ounce of talent, and the people at TPI stand on the firm ground of science. Every question has an answer, and no mystery can't be solved.

So Lance and Sophia went through a checklist of concrete goals and supplemented it with a lot of targeted affirmation and positivity. When he thought she was ready, he lightly tapped his desk with his hands and stood up. "Let's do it," he said, then they walked outside to the practice range.

When Dave Phillips, one of the cofounders of TPI, heard about Sophia's tests and her coach's response, he let out an abbreviated smile.

"The teacher in golf," Phillips said, "has evolved."

In the early 2000s, when Phillips was a renowned teacher working at a club in Maryland, he had a talented teenage student named Peter Uihlein. Peter was the son of Wally Uihlein, the longtime CEO of Acushnet, then the parent company for Titleist. Peter would go on to win the U.S. Amateur in 2010 and turn pro soon thereafter. Long before then, Phillips was already thinking big about the future of golf instruction. He called Wally to come down to the Washington, DC, area, where Phillips had developed a relationship with Dr. Greg Rose, a physical trainer and chiropractor. Rose owned a budding business called Club Golf, where he was making his name working with a lot of long-drive champions. With Phillips's background as an instructor and Rose's scientific outlook on the body, the two were already far out in front of the trend concerning biomechanics and the golf swing.

"That's the future of golf instruction," Phillips remembered Wally Uihlein saying after he saw how the two worked together. "We [Titleist] need to figure out how to be a part of it."

Wally Uihlein recognized almost immediately that this type of instruction could exponentially expand his business—and brand.

Titleist had always based their business model on owning a majority share of the golf-ball market. They estimated that the average golfer lost approximately six balls per round. The more rounds a golfer played, the more golf balls he bought, and the more money Titleist stood to make. As players got better, the golf balls they purchased grew more expensive. Better golfers may lose fewer balls, but the market for high-end balls was utterly dominated by Titleist. That was especially true after the sensational release of the Pro V1, when forty-seven PGA Tour players put the prototype in play for the 2000 Invensys Classic in Las Vegas, likely the largest full-scale equipment shift in the history of the game. Once the ball shipped publicly in March 2001, it remained the bestselling ball (along with its later offshoot, the Pro V1x) for close to the next two decades.

So making players better, finding new areas for technical improvement, and allowing people to play longer—it was all about sustaining profit for Titleist through the sale of more golf balls.

Listening to his son's golf coach talk about the science-rich future of golf instruction, Wally Uihlein found a way to inflate his biggest market advantage. Publicly investing in performance science made it look as if the goal of Titleist was mainly to be at the cutting edge of technology. Titleist could market the golf ball without the consumer realizing that anything was being sold.

"It's pretty smart for our CEO to sit there and go, 'We think we should be looking at every aspect of golf, from the physical side to the mental/emotional side—everything to have a golfer love the game more, hit the ball farther, and enjoy the game more,'" Phillips said. "Because if they go out and enjoy the game more, chances are they're going to be playing our golf ball."

Phillips speaks with a clipped accent that is hard to place at first. It's mostly from his parents, who were both from England, where he was born. When he was six months old, his father, working in telecommunications for the British military, moved the family to Kenya. From there, they moved all over Africa, then to the Middle East and the Far East, finally settling in Australia when Dave was in his teens. With tightly cropped and receding hair, an angular face, and deep-set dark eyes, he gives off an aura of weathered worldliness.

Throughout his teaching career, Phillips had always been bothered that no matter how good the instructor, some students improved and others didn't. It bothered him in the mid-1990s when he worked for David Leadbetter, using Phillips's own proprietary video software called NEAT (Never Ending Athletic Trainer) to tape eight-hour practice sessions with Nick Faldo during his prime. It still bothered him in 2000, when he was thirty-two years old and became the youngest ever to be named to *Golf Digest*'s Top 100 Teachers list. The magazine wanted to do a story comparing him and Paul Runyan, who was the oldest on the list at ninety-two, but Runyan died before they could get around to it.

When Phillips met Rose and saw the kind of work he was doing, Phillips finally began to hope instead of despair. Rose's methods were preemptive, beginning with putting every golfer through a "physical screening" to identify his or her limitations before finding a methodology that might work. Phillips first came to Rose with a low-handicap student who was struggling to implement what was being taught. Right in front of Phillips, the student was put through about five minutes of stretching and strength tests, while Rose took notes. Tall and muscular, handsome with a neat part in his wavy light-brown hair, Rose is undeniably charismatic. He speaks with a smooth authority and leaves little room for argument. At the end of the student's evaluation, Rose handed his notes to Phillips and said, "That's what he's going to do in his golf swing, and if you try to do anything else, you're going to struggle."

Phillips laughed when he remembered the story because it was a life-changing moment for him. Rose had written down roughly what was happening with the student. Phillips had struggled to get the student to complete his turn on his backswing, then struggled to get his hips to open up through contact and get his weight to the left (front) leg. As a result, the student often hit a thin shot to the right (or a dead chunk to the left) when under pressure. What Rose had gleaned from the evaluation was that the student had little flexibility in his hips, making it difficult to fully turn; a lack of strength in his left knee (from an old injury) that kept him hesitant to shift hard to his left side; and a slight lower-back problem that had developed from practicing so much without the flexibility needed to execute the instructions Phillips was giving him. Rose physically described *why* this student wasn't improving with what Phillips was telling him.

"It was like, you get these moments, like the bright light went off in my head," Phillips said. "It was like, 'That's it! That's the reason why all these [teaching pros] struggled with some and were successful with others.' It wasn't that [Leadbetter] or any of these other great teachers were trying to be bad. They had great techniques. It's just that they didn't know because we were never taught. We were golf pros, we weren't taught about the body and how it works. No one had taught you a simple way of evaluating the body so you could understand why you were different than me. Right?

"So that, to me"—Phillips threw his pen on the desk—"that's it! That's the thing!"

In addition to simple physical assessments, Rose was also an expert in 3-D motion analysis. His system, which would eventually be installed at TPI (and would later be replaced by something more sophisticated), was centered around little computing sensors, slightly smaller than Ping-Pong balls, that were attached to clothing. The outfit consisted of a hat, a vest with shoulder pads, sheaths for arms and elbows, gloves, a garter-like piece for the hips, braces for the knees, and thin covers for the shoes. So dressed, a test subject would practice in a room that was set up with special receiving cameras all over the walls and ceiling to pick up the exact location of each sensor. Sometimes a special golf club that carried smaller sensors at the grip and clubhead was used, as well. The motion analysis was similar to what computer programmers used to develop video games.

In real time, an animated stick-figure version of the player would appear on a computer screen. After recording, the movements of the figure could be played back and forth in slow motion. More important, the computer calibrated numerical data about the physical relationships between body parts as they were all in motion. The result was a detailed elimination of proprioceptive dysfunction.

Such technology could collect large amounts of data about the body and its mechanics, but most frequently focused on was a pattern of movement that all great golfers have followed, even if their swing paths were aesthetic opposites. One way the data was manifested was on a graph, with body and club rotation on the y-axis (vertical) and time elapsed going left to right on the x-axis (horizontal). The different body parts were identified by different-colored lines, so as the player started the backswing, the lines would move from left to right (along time) and dip below the equator in varying degrees of curved parabolas depicting the amount of rotation back. Then the lines curved up and started to rise when rotation slowed down, and then overlapped at the equator, showing the top of the swing when rotation started to move in the other direction. The lines then crossed the equator and hit a peak height at impact before slowly descending again. So down, then up, then down again, like a bunch of multicolor sideways *S*'s that varied from fat to skinny depending on the amount of rotation with each body part.

What teachers were looking for—and what the graph made easy to understand—was any deviation in the kinematic sequence that all great golfers have found intuitive. First the hands move, then the arms, shoulders, hips, knees, and feet. The sequence goes in reverse on the way down: feet move first, then knees, hips, shoulders, arms, and hands. The data from the 3-D motion sensors can be extrapolated to create a separate graph of each body part being monitored, with a more detailed analysis of the movements, making the kinetic and sequential comparisons more specific than the original graph. What looks good on the big graph might have small variance depicted in the smaller graphs.

Overall, 3-D motion tracking allowed experts to further analyze the swing in a deeper, more concrete way than with just video and the naked eye. The live swings of Jim Furyk and Ernie Els could hardly look more different, yet their data-plotted graphs are almost identical. Leadbetter called it "syncing." With the graphs, it was easy to show a student how his or her hips stopped rotating before impact, and how that threw off the rest of the sequence, likely forcing the arms to get out in front. The lines would get all jumbled and the arcs wouldn't coincide. Such technical language didn't always make sense by itself, such as in trying to explain how and when the rhythm went askew. Tempo is an abstract, but it becomes far more tangible when illustrated with colors and charts.

Rose was exceptional at interpreting this information and disseminating what it meant to the big picture of a person's golf swing. If the graph showed a lack of pelvic rotation, it might be due to a lack of flexibility in the hips, which Rose would have detected in a physical screening. If the hands stopped rotating, it might be due to an old injury in the player's wrist. The information being collected from physical screenings and 3-D motion analysis better explained exactly what was happening in the golf swing, and Rose then used it to explain *why* people moved in a certain way—and why that didn't always coincide with what they were trying to do, or what their teacher was saying.

With Phillips's background in technical swing mechanics, the two brought the whole process of improving as a golfer into clearer focus. They could now understand what swing motions might work and not work depending on physical limitations and understand how to keep a player from suffering an injury. They could now tailor the golf swing and any possible improvements to each person's individual biomechanical makeup. This targeted instruction was exactly what Phillips had been searching for, and he knew it when he left Club Golf that first day.

By 2003, Wally Uihlein had convinced the two to take over an underutilized club-testing facility in Oceanside, and they named it the Titleist Performance Institute. They spared no expense on the grass practice areas and the gym. All of the 3-D motion analysis equipment was set up in an enclosed bay, and all the tech was upgraded whenever possible.

"Our passion and our desire back then was to become the Mayo Clinic of golf," Phillips said. "We wanted to look at everything that affected a golfer's performance—other than equipment, because we already did that—and see whether we could find anything."

They began by bringing in PGA and LPGA Tour players who were sponsored by Titleist. They ran through exercises on the early ball-flight monitors, suited up in 3-D, tested new equipment, and went to the gym with Rose. TPI eventually expanded to "fantasy camps," as Phillips called them, expensive one-to-three-day experiences where the public could come in and get the same Tour-level treatment. But Phillips and Rose quickly realized that they could only see so many people. If TPI was truly going to follow Uihlein's dream business model, they needed to get the word out to the public.

So TPI started to "certify" people in three categories—instruction professionals, fitness professionals, and medical professionals. Each one had its own subsections, such as teaching juniors or physical therapy or nutrition, and each had its own levels of advancement. These newly certified professionals would be the disciples set to spread the word.

TPI quickly grew so big that they starting booking the entire convention center of the sprawling JW Marriott in Orlando for their annual World Golf Fitness Summit. Events such as this had become staples of the TPI business plan. The summit offered a mix of presentations from top-ranked teachers such as Sean Foley and Mike Adams, followed by individual classes taught on specific subjects. Hundreds of people attended the 2012 summit, paying between $100 and $600, depending on what type of certification they were seeking or if they were just attending as spectators. Many attendees wore headsets during the presentations, with a table of interpreters in the back to translate. The hallways were lined with booths from manufacturers pitching their new products, all paying TPI just to get that type of exposure.

In the morning, after Phillips and Rose spoke and after one or two presentations from visitors, about two hundred people looking for basic Level I certification headed off into a grand ballroom. This was the core class where all TPI certifications began. On the stage was a highly ranked instructor named Mark Blackburn. Wearing a headset microphone and displaying the look and energy of an infomercial salesman, Blackburn began to explain to those on the floor how to administer an initial physical screening—the screening Rose had performed so many years ago. As Blackburn demonstrated, the participants began to practice on one another.

Using a line of tape laid down on the carpeted floor, and a thin,

three-foot wooden pole, they were told to measure how far their partner could stretch, to ask them about any previous injuries, and to
record the results on a TPI-produced template form. After the exercise
was over, Blackburn gave a presentation on elementary swing mechanics. Some people in the audience had volunteered tapes of their
swings, and Blackburn called on them to stand up. One man's setup
position was projected onto a large movie screen behind the stage,
and Blackburn drew all the standard lines of shaft angle, head position, and knee bend, the result looking very much like a diagram out
of Phillips's NEAT program from the late 1990s. Blackburn identified
in the man's form some discrepancies to what could be considered
Leadbetter's mechanical orthodoxy and described some exercises that
the man could do to increase his ability to execute better positions.
Blackburn also mentioned that the exercises would improve the player's proprioception. Explaining that fascinating concept elicited no
perceptible excitement from the sleepy congregation. A man in the
back, wearing stained sweatpants and a Florida State T-shirt, stopped
listening and stood up to start practicing his swing. For about an
hour, Blackburn's voice reverberated around the room like the dron-
ing hum of a washing machine. When it was over, these people in the
audience became TPI-certified Level I instructors.

"Remember," Blackburn said near the end, "when you leave here,
you're the experts."

In another room, after a boxed lunch of tuna salad and sliced
apples, a man named James Sieckmann gave an indoor chipping lesson.
Phillips described Sieckmann as "the best wedge teacher in the world,"
and his session was limited to a more concentrated group of about
thirty people, already at Instructor Level III. Sieckmann had taught
many PGA Tour pros, and in the small conference room he was canon-
izing the chipping technique of two-time Tour winner Tom Pernice Jr.

"Throughout the history of the game," Sieckmann said, as video
of Pernice hitting a 20-yard chip played on loop on a projection

screen, "the best finesse players are not the best ball-strikers." A couple of students then hit pitch shots off the carpeted floor with soft plastic balls. Sieckmann made a few small adjustments with setup and swing positions—then asked each student to do the same to another student—and everyone clapped.

By the time of the 2012 summit, Phillips estimated that close to thirteen thousand people in fifty-five countries were certified TPI instructors, and many more uncertified teachers and players were out there using TPI's principles.

"If you think about each one of those people and how many golfers they teach or affect," Phillips said, "you can start to see the staggering amount of people we have affected over the years."

TaylorMade has always had a business model based on their large share of the golf-club market. Whereas Titleist wanted players to improve so they kept buying top-of-the-line golf balls, TaylorMade wanted players to improve by using their clubs. This way, golfers felt aligned with the brand and kept coming back for upgrades. The technology TaylorMade utilized served their purpose.

The company was so confident in their research and development team, and so confident that their clubs were the best, that they focused on finding and fine-tuning technology to prove that point. That led to a relationship with a company called Motion Reality Inc., which had collaborated with the U.S. military on projects, in addition to working with the gaming world to develop realistic computer graphics. By 2002, TaylorMade and Motion Reality Inc. had collaborated to develop a system for golf named MATS (Motion Analysis TaylorMade System, later to be known at MAT-T and then Gears), similar to the 3-D sensors used at TPI. MATS was not necessarily used to improve a player's golf swing, but to make sure he or she was fitted for the set of TaylorMade clubs that could help the most.

The company already had their own version of the TPI campus, dubbed the Kingdom. Located in Carlsbad, less than eight miles south of TPI, the Kingdom was as regal as it sounds. On one side of the street was the corporate headquarters, a sprawling complex of three-story buildings that included the cubicle offices for marketing and sales, along with the labs for R&D, the latter located behind a large door with the sign OFF LIMITS. All the clubs were assembled in the huge warehouse, with the parts made in Asia. The complex was home to about a thousand employees, with a cafeteria and some seating under umbrellas outside. It could have been a nice corporate park anywhere in America.

But across the street, a Spanish-style stucco building sat at the top of a small hill, behind pristine, mossy landscaping that gave it the feel of a European country villa. A large wrought-iron fence ran around the sides, and thick wood shutters covered the windows. Inside, the lobby was all dark wood and dim lighting. Hanging on one wall was the TaylorMade driver that Retief Goosen used to win the 2004 U.S. Open at Shinnecock Hills, as well as the hybrid that Y. E. Yang used to hit the deciding shot to beat Tiger Woods in the 2009 PGA Championship at Hazeltine. Toward the back of the building, a glassed-off room had some machines and AstroTurf used it for putter fittings. Out back, an expanse of green grass opened onto the driving range.

The MATS system was in a big, enclosed room with six cameras posted on the walls, and six sensors at different points on each club, with a selection of driver, 6-iron, and wedge, all to be used in a fitting. The person to be fitted would suit up from head to toe in pieces of black nylon, secured with Velcro, containing twenty-eight sensors in all. With one swing, a plethora of graphs and numbers popped out onto a computer screen and were archived. They eventually created a swing profile. Rather than being used to give a lesson, the numbers were used to match the player with the proper set of TaylorMade clubs.

"TaylorMade knew they had an unbelievable system when this came out, but right then, [the MATS system] wasn't their focus," said Tom Fisher, who got his hands on a MATS that first year, 2002, when he was working at the Belfry in England. "Back then, their focus was getting the number one driver, the number one irons, and world domination in terms of golf-club market share. And they did."

Fisher played college golf in England, and when he got to the Belfry, outside Birmingham, it was already a stronghold for TaylorMade, one of the company's most productive club sellers in all of Europe. TaylorMade came to an agreement with an investor to commercialize the MATS system—a wealthy investor who went to the Kingdom for a fitting and fell in love with the technology. They would scale the system down a bit and turn it into the TaylorMade Performance Labs, bringing them all over the world. They would train the club fitters how to use the system and watch as TaylorMade clubs flew off the shelves. The company made the commercialization agreement with the investor with the provision that TaylorMade would oversee each expansion and would have a say in who was hired and how people were trained. They opened seven labs in four or five years, then, per the agreement, the company bought the system back in 2010.

One system landed at the Belfry, where Fisher embraced it. For each of the next three years, he either doubled or tripled sales of TaylorMade equipment. He also started to get restless, and in 2005 he got wind of a possible opening in California at a Performance Lab based out of Aviara Golf Club and Resort, just over a small mountain from the Kingdom—"about five drivers away," Fisher said. In his first year there, Fisher said he did $1.3 million in club sales.

In March of 2012, he moved over to be the business development manager at the corporate headquarters and watched as the rise in technology helped to improve profits in the sale of his golf clubs. If the average player gets better in the process, then so be it.

"We train our fitters, but they're not teachers, and we make that abundantly clear," Fisher said. "But you have to come from a swing-analysis background in order to communicate how the clubs work."

It makes sense that TaylorMade would be the company to focus on golf clubs, considering that's where it all started in 1979 when an Illinois golf-supply salesman named Gary Adams took out a $24,000 loan against his house and started the company with one patent—the cast for a twelve-degree metal driver. The son of a teaching professional back in McHenry, Illinois, Adams had been dumbfounded how the newer balata balls were flying farther off the face of irons, but not the persimmon woods. The club he designed was radical not only in its material, but also, because of the pliability of metal casting, he could control how the weight was distributed. This new metal driver had more weight around the perimeter of the club so it could remain stable through mishits, and it also had a lower center of gravity to get the ball into the air quicker. Both of those ideas would become tenets in golf-club design going forward.

Adams brought a trunkload of his metal drivers down to the PGA Merchandise Show and started selling them to club pros to stock in their shops. That first year saw about $47,000 in sales. By the time Jim Simons used the club to become the first winner of a PGA Tour event with a metal wood at the 1982 Bing Crosby National Pro-Am, the death knell for persimmon rang like one loud *ping!*

By 1984, Adams sold the company to the ski maker Salomon, which was eventually bought by the German apparel company Adidas in 1997. TaylorMade continued to innovate and kept control of the club market with the Burner Bubble series in the late 1990s. Many more innovations followed, and by 2006 they had become the first golf company since Acushnet to create $1 billion in revenue.

Just as Titleist's brand was created in part by its being the "No. 1 ball on the PGA Tour," so did TaylorMade push for most of the 2010s that they were the No. 1 driver on the Tour. (Callaway beat them out in 2017.) When Nike stopped making golf equipment in

2016, TaylorMade was first in line to pick up Tiger Woods and Rory McIlroy for equipment endorsements, again striving to show that their clubs were the best, trusted by two of the best players of their respective generations.

But TaylorMade's fitting technology was no longer proprietary, and most golfers lived within a reasonable distance of a place that could do advanced club-fitting. Turns out these places carried more than TaylorMade, as well. But that didn't stop the company from trying to innovate further, experimenting to see how technology could be used to take advantage of their biggest market share.

"I get this all the time," Fisher said. "A fifteen-handicapper who is slicing it comes in, and I give him clubs that stop it, and two years later he's a five-handicap and he's upset he needs new clubs. I say, 'Well, I'm really sorry that these clubs took ten shots off your game. That's such a bind. How about we just get you new ones?'"

Justin Thomas is a Titleist guy, starting from when he was a star at the University of Alabama and continuing when he signed an endorsement deal with the company once he turned professional in 2013. In August 2017, the diminutive Kentuckian with a booming golf game walked up to the driving range at Glen Oaks Golf Club on Long Island, preparing for the first leg of the FedEx Cup playoffs. He had just won his first major championship earlier that summer, taking the PGA Championship at Quail Hollow Club in North Carolina. He would eventually win the FedEx Cup and collect his first PGA Tour Player of the Year award.

But on this quiet Tuesday on Long Island, Thomas's Titleist staff bag was being carried by his caddie while Thomas himself was carrying a small padded satchel, gray with orange trim. Inside, about the size of a small briefcase, was his personal TrackMan, the newest version of the machine, which had cost him, with the Tour-pro discount, somewhere just south of $20,000. He set it up behind his hitting station, and it created a Wi-Fi network that he left public, making it available to connect to if you had the TrackMan application for a smartphone. The pro at Glen Oaks, Tim Shifflett, opened the app, connected, and turned to a nearby spectator to show that Thomas just flew a Titleist 3-wood 294 yards in the air, with an optimal minus-3 attack angle and somewhere around a 3,600 rpm spin rate.

No words were exchanged, and none were needed. Nor did any numbers need to be crunched to understand that Thomas had absolutely crushed it.

As you looked down the driving range that day, 90 percent of the players had their own personal TrackMan set up, and they would hit a shot and walk back to either a caddie or an instructor holding an iPhone or iPad to look at the numbers. Even Dustin Johnson, with his unconventional swing and his laissez-faire attitude, credited his work with TrackMan for dialing in his wedge game to an extreme exactitude that eventually led to his first major championship at the 2016 U.S. Open at Oakmont.

To see these tools ubiquitous on the range at the highest level of the game was the genius of the TrackMan business model—they made everyone think he or she needed one. From instructors to players, the belief proliferated that you had to understand how these numbers worked or you couldn't fully understand the game.

"A lot of times in this game, what we're feeling is not exactly what we're doing," Tiger Woods said in 2011, a year after beginning to work with Sean Foley. "I just think that you're trying to match up 'feel' and 'real.' And as you make swing changes, you make slight alterations, you start realizing what [the club] does at impact, and what that can translate into in the performance of a golf ball."

Justin Rose sent Foley an email on the eve of the 2013 U.S. Open at Merion with a copy of his TrackMan numbers from a practice session at home before leaving for Pennsylvania. Foley wrote back, "Rosie, awesome, love those numbers, can't wait to see you tomorrow." Rose went on to win that U.S. Open, flushing a 4-iron on the seventy-second hole just next to a plaque that commemorated the historic 1-iron hit by Ben Hogan in the fourth round of the 1950 U.S. Open.

"Is it transformational? I think it is if you understand how to do it," Woods said. "But also not to get embedded in it where you start losing your feel and your touch."

TrackMan was born by hitting golf balls off a U.S. warship.

Starting in the early 2000s, a Danish golfer-turned-doctor named Klaus Eldrup-Jørgensen teamed up with his brother, Morten, a successful lawyer in Copenhagen, to start investing in driving ranges across Europe. They understood that technology would eventually get to a point where people could to go a driving range and get all the information about their shot that they wanted—how far it went, ball speed, club speed, etc. Rather than watch someone else cash in, the brothers wanted to be the innovators.

They found that one of the biggest contractors for government defense departments—including the United States—was a Danish company, Weibel Scientific. They dealt in such things as missile tracking, so maybe they could help with his little problem of tracking a golf ball. They set up a meeting, and Weibel politely told them they didn't have the time for something of this stature. Yet their number one engineer, Fredrik Tuxen, was intrigued. He took the brothers aside after the meeting and told them that the technology was certainly possible. He knew it for a fact.

When testing radars on U.S. warships, Tuxen and his crew used to hit golf balls off the deck to make sure the signals were being picked up. If the radar could pick up the golf ball, it could certainly pick up a missile. None of the specifics of the ball were being measured besides its path, but the rest of the data was there to be disseminated.

The trio formed TrackMan in May of 2003, spending the first year refining the product out of Klaus's garage, trying to ensure their machine could measure the ball in the air from start to finish. They got the product down to about the size of a phone book, meant to be placed behind the golfer and down the target line. Many ball-flight monitors were on the market, some camera based, some using lasers. To some extent, those all guessed. TrackMan saw. When ready, the men of this new company called up all the big golf manufacturers in the United States and asked to come in and show their product. Almost all the R&D people thought that TrackMan couldn't possibly do what they said it could, but they invited the Danes anyway.

TrackMan had five meetings set up on that first trip stateside, and its engineers gave demos and presentations, explaining how this small device could retrieve all this information instantaneously. There was no more guessing. This technology would save the companies time and money and make their data more exact. The R&D people were floored. The TrackMan team went back to Denmark with five contracts for purchase, each system costing somewhere near $100,000.

At one of those meetings, a R&D person asked if TrackMan could also track spin. Tuxen, the lead engineer, was not much of a golfer at the time, having only flailed at those few balls off the deck of a ship. He didn't know that tracking spin would be something of interest. So he went back and found a way to pull the signature of the spin out of the signal and produced the spin-rate number. By mid-2007, the machine could produce numbers defining the swing itself, such as club path and attack angle and face angle at impact. It tracked the club from its center of gravity, not the face, and the contact point on the face remained a mystery. But TrackMan's numbers were eye-opening.

The most shocking truth, from the indisputable evidence, was that the starting line of the ball is almost entirely determined by where the face is pointing. For decades, professionals taught that the ball started on the path of the club, not from where the face was pointing. That was even written in some old PGA of America teaching pamphlets. The professionals were dead wrong, and when TrackMan salesmen came around to show them that, more than a few disagreed.

"There was a lot of kickback there, that 'Oh, it's not accurate. How can you prove that?'" said Justin Padjen, who joined the company in 2006 as a salesman and business development manager based in the United States, often traveling on the PGA Tour. "I wouldn't say that occurs anymore."

Although the TrackMan systems started showing up more frequently, they were still primarily thought of as technical measuring devices for manufacturers and occasionally used by high-end club fitters. Padjen estimates that focus changed sometime around 2010, when teachers started to embrace how the data could help their students.

David Leadbetter bought one of the first ones, available sometime around 2006, and Sean Foley bought one a couple years later. "It was like a homemade bomb," Foley told *Golf Digest* in 2014, describing how he used to keep the TrackMan website ready to pull up for TSA agents when he was bringing it through airports.

Some players became overly obsessed with the numbers, none a more cautionary tale than Trevor Immelman. In 2008, Immelman was a tidy player out of South Africa with a beautiful-looking golf swing. That year, he held off none other than Woods at Augusta National to win the Masters for his first major championship. From there, his internal expectations changed. Twenty-eight years old, he thought his game needed to get a lot better, and to do that, he needed to hit the ball higher. He started working with Claude Harmon III, Butch's son and a member of the advisory board at the Titleist Performance Institute. Unlike his father, Claude was deeply reliant on technology, and he used TrackMan constantly. On TrackMan, Immelman saw that the path of his driver at impact was two degrees down. Ideally, you want to hit up and out on the driver, not down. Immelman thought that was the problem, and he tried to fix it.

"You won the Masters," Claude remembered telling him. "You don't need to hit it higher."

Immelman injured his left arm, getting tendinitis in his elbow and wrist. By the end of 2013, he had split from Harmon and still hadn't won another tournament, forced to go play on the Web.com Tour Finals to regain his PGA Tour card, which he did.

But stories like that don't daunt the instructors who have made a name through technology. Grant Waite was a good player out of New Zealand who won the 1993 Kemper Open. He was also famously the runner-up to Woods at the 2000 Canadian Open, when Woods aimed at Waite's ball on the seventy-second hole and hit one of his most iconic shots, out of a fairway bunker, over water, to about eight feet. Waite's game sagged, and he became obsessed with the mechanics of the golf swing. He went down a rabbit hole, with *The Golfing Machine* at the bottom. As he struggled to make a living on the Tour, he started giving lessons, and when TrackMan came around, he fell head over heels. He hooked up with another teacher of the same inclination, Joseph Mayo, who was the director of instruction at TPC Summerlin in Las Vegas. Mayo would soon gain a large Internet following, using his popular Twitter account, @TrackmanMaestro, to argue over the truths of the golf swing.

"There's no way that you can make the statement that player 'X' can't reach his potential without technology 'Y.' You can't make that statement," Mayo told Golf.com in 2016. "But what I will say is this: If a player has all the technology available to him, which means, in a way, you could say he has more answers, he has more information, then I think every player could improve.

"Because we've had the Jack Nicklauses, the Arnold Palmers, the Sam Sneads—they didn't have that technology, and that's an argument that I get all the time. 'Hey, Ben Hogan didn't have TrackMan.'

Or, 'Jack Nicklaus didn't have TrackMan.' And my rebuttal to that is this: Those are the most gifted athletes that have ever played the game. And they instinctually discovered a technique that worked for them. But 'John Smith' down here on the range, he doesn't have that instinctual ability."

That is where Mayo thinks his work as a teacher is proprietary. To be able to use TrackMan, in concert with the most high-end 3-D motion analysis and an understanding of biomechanics that comes with that, he can give a more definitive lesson than ever before. That rubs many the wrong way.

"It seems that there is a group of instructors out there that fear technicality," Mayo said. "They fear knowledge. And I don't agree with people getting too technical. I think that everybody learns differently. No question about that. But fear of knowledge, and fear of technical expertise, is not the answer."

Padjen said that in 2015 over 350 touring pros, from the PGA, European, Asian, and LPGA Tours owned their own TrackMan system. Unlike any other piece of equipment they use, they all had to reach deep into their pocket to get one, getting it discounted to $17,500 only if they allow TrackMan to use their numbers and likeness in their worldwide programs. Striving solely to reproduce those numbers, talented amateur players could make some strange movements to achieve the same results, but could also end up with a swing that is physically unsustainable. Some teachers call that "fudging" the numbers. Other players find that by using TrackMan properly they can bypass the teacher altogether.

"It's been a process," Padjen said, "because looking back at my early days, I would say there was a lot of fear out there. But TrackMan isn't taking anyone's job."

Then there's this, from Foley in *Golf Digest*:

"If I took a kid who is a really great player, really skilled, twelve years of age, and I taught him everything about TrackMan until he was fifteen, he'd never need a coach. If I said, 'Look, this is how you move it to the right, this is how you move it to the left. If it gets too steep, this is how you get it out of the ground, this is how you can hit it farther, this is how you can hit it lower,' then he would just be able to find it in the dirt the way Ben Hogan did. And all the learning experts are saying that's where true learning is done. From just doing it. Trial and error. TrackMan tears down the method-and-model instruction."

The PGA Tour is a business, one that's mostly predicated on getting people interested in their television broadcasts so the Tour can make money on advertising and sponsorships. To make the telecasts better, they eventually embraced TrackMan, which shows the flight of the ball in the air and makes it far easier to know what's happening. But sports fans also like easy-to-understand statistics, and the Tour always struggled to have anything more insightful than fairways hit or greens in regulation.

It was not too different from baseball, likely the closest sport in being hunkered down by tradition. And just like baseball, golf is a slow game, with each piece of action having a stop and start. Hence, both are easily chronicled and can amass huge amounts of data.

The revolution in how to use that data in baseball was instigated by Billy Beane, general manager of the Oakland Athletics and star of Michael Lewis's 2003 book, *Moneyball*, later to be a successful movie staring Brad Pitt as Beane. The Athletics hired young, smart businesspeople to figure out what actually won baseball games, and how to evaluate players in that context. Despite a meager payroll, they were successful, and the new statistics they had helped develop eventually became used by all the clubs.

The PGA Tour had a lot of data as well, especially after ShotLink was used, starting in earnest in 2003, to chronicle the details of every shot hit by every player. Mark Broadie was desperate for access to that information, eager to analyze it in the mathematical way he analyzed the business world while part of the faculty at Columbia Business School in New York. With a doctorate in engineering from Stanford, Broadie did research into quantitative finance and published many academic papers while teaching a class called business analytics. He was applying math to the business world, doing such things as options pricing and derivatives. He then picked up golf again after toying with it throughout his childhood in southern New Jersey and his undergrad studies at Cornell.

A friend at Columbia that he often played with was on the handicap research team of the United States Golf Association (USGA) and convinced Broadie to join up. Once Broadie started analyzing the processes for how to develop course ratings and how to adjust golfers' handicaps, he saw a connection between what he did during the week and what he did for fun on the weekends.

"The questions I was interested in asking were things like, If an amateur could hit the ball twenty yards farther, how much would his score go down?" Broadie said. "Or, where did the ten shots that separate a ninety golfer from an eighty golfer come from? Or the strokes between an eighty golfer and a club pro? Or an eighty golfer and a PGA Tour pro? Where exactly are all these strokes?"

By 2005, Broadie had developed a software program called Golf Metrics, and he called the PGA Tour, asking if he could have access to all of their data to make an academic analysis. They balked. The Tour had paid a lot of money to figure out how to track all this stuff—including purchasing TrackMan systems—and they didn't want to give it away for free.

So for the next two years, Broadie would call up, ask the same question, and get the same answer. But come 2007, the Tour flipped the script. They told him that they had all of this data, and, golly, they just didn't know what to do with it. What the Tour now recognized was that they needed to find some valuable statistics before someone else did. Today's average sports fan longed for a depth of understanding that could only be mined from the new reams of data. To make golf more interesting—and therefore to make more money—the Tour wanted better statistical analysis.

In 2008, Broadie started publishing his findings. What he created was a new statistic that he thought was more accurate than the standard statistics the Tour was using, and he named it *shot value*. Soon realizing that was an established term used by golf-course raters for magazines, Broadie changed the name to *strokes gained*.

For every shot hit, Broadie came up with a Tour average result—for that day, that week, and eventually that season and then groups of seasons. The data made it possible to compare players' performances against the field.

What the comparative data showed was entirely different from the traditional way people thought about the game. The long-standing idea that the short game is what matters most in scoring was dead wrong. In comparing where good players had the biggest advantage, Broadie found that two-thirds of their "strokes gained" were from outside one hundred yards.

"The more I thought about it, and dealing with the data constantly, the more and more it made sense, and it became completely obvious that this is the right answer," he said. "It wasn't obvious at the start, and it's still not obvious to many people."

Here's an easy way Broadie explains it to people, and it's a brief summary of his 2014 book, *Every Shot Counts*. Pick your favorite PGA Tour pro and decide if you want to have an eighteen-hole putting competition with him, or an eighteen-hole driving competition with him. The putting competition you might have a chance at winning; the driving competition, zero chance. How about a competition for proximity to the hole on par 3s between 180 and 220 yards? How many scratch golfers would even have a chance against a middling Tour pro? None.

When the data-supported conclusion is laid out like that in black and white, it surely seems logical. It's almost impossible to think it took people this long to realize it.

"It's sort of like the horse-racing analogy—you can't win a horse race in the first 200 yards," Broadie said. "A lot of times it comes down to the stretch run."

The stretch run in golf is the putting green, where so many tournaments seem to be decided. Between two players? Yes. Between them and the rest of the field? Not really.

Broadie was quick to point to the 2015 U.S. Open at Chambers Bay. Coming to the seventy-second hole, Jordan Spieth and Dustin Johnson were in different groups, but were virtually tied atop the leaderboard. Spieth was on the green of the par-5 eighteenth hole in two, and two-putted for a birdie. Johnson was on in two, as well, and was even closer, but three-putted and handed the tournament to Spieth.

"It's pretty easy to conclude—I think wrongly—that Dustin Johnson lost the U.S. Open because of his putting, or Jordan Spieth won because of his putting," Broadie said. "What that misses is how did they get to that position on the eighteenth hole in the first place?

"If you have two people that are equally good from tee-to-green, then it can come down to putting. But you have to be great from tee-to-green for putting to matter. If you're a great putter, that's not good enough."

Not only has this system found a way to analyze the game as a whole, but it can break down each segment of the game and give a better evaluation of a player's performance. For putting, if the Tour average is exactly two putts from thirty-three feet, then a player who makes a 33-foot putt has gained one stroke on the field. If the Tour average is one and a half putts from eight feet, then if a player makes an 8-footer, he's gained half a stroke. By using the comparison method and some simple subtraction, the system has eliminated some misconceptions that are held in traditional putting stats.

Take Johnson, for example. The week after the 2015 U.S. Open, the athletic young bomber was ranked tied for 13th in putts per round (28.20) and 2nd—*2nd!*—in putts per green in regulation (1.714). Seemed pretty good. In strokes gained-putting, he was ranked 125th, losing .066 strokes per round to the field, an evaluation that seemed a lot more commensurate with his performance.

"Night and day—they can't both be correct," Broadie said. "It's pretty clear he is a below-average putter, and strokes gained-putting nails it correctly. And putts per round and putts per green in regulation get it horribly wrong."

When the Tour saw these types of results from the work Broadie was doing, they knew they couldn't avoid it. They got the word out to some influential writers and magazines, and by 2010 they had slowly rolled out "strokes gained-putting" for their own website and publication. Yet they remained cognizant of not overwhelming golf fans with statistics that they didn't understand.

In September of 2014, they brought out "strokes gained: tee-to-green." The plan was to soon thereafter break that down into three categories—around the green, approach shots, and drives, all to go with the established putting stat. Because there is now data for every type of shot, it can be sliced as thin as one likes. Broadie said increments of less than 50 yards—anything smaller than 50 to 100 yards, or 100 to 150 yards—can limit the sample size and create weird and unreliable results. But coaches do want more specifics than are available on the Tour's website, and Broadie is willing to provide them.

"So much of what we believe is handed down through nostalgia," Foley told PGATour.com in March of 2014, just after he and Broadie put together a joint presentation at MIT's annual Sloan Sports Analytics Conference.

Foley told a story about how he had been on a plane home from a tournament with Justin Rose in 2013, and he asked Rose how he felt about his wedge game. Rose replied that he thought his wedge game was okay, but he wanted to be better; he wanted to be like Luke Donald and the other great wedge players in the game. Then Foley showed him that for 2012, Rose was No. 1 in the world in strokes gained from one hundred yards and in, and Rose was shocked.

"What happens to athletes, because they're always in the thick of it," Foley said, "they start telling themselves stories."

A lot of the golf world has come to believe that technology has made the game easier to understand by producing far more information, and far more concrete analysis. And all the while, Titleist is selling golf balls, TaylorMade is selling golf clubs, TrackMan is selling expensive ball-flight monitors, and the PGA Tour is selling their game through new age statistics.

Turns out, the necessity for profit remains the mother of invention.

# SEVEN

# THE ART OF ARCHITECTURE

By 1960, the world's oldest and most prestigious golf tournament had become an afterthought for American professionals.

In the two decades leading up to that Open Championship at St. Andrews in Scotland, the number of Americans participating in the tournament had dwindled. Gone were the days of Gene Sarazen, Walter Hagen, Bobby Jones, and Tommy Armour going over every year. Americans had actually won the tournament for ten straight years from 1924 to 1933. But by the time Ben Hogan finally went over and completed the career Grand Slam with a victory at Carnoustie in 1953, he was one of only four Americans in the field. When Sam Snead won in 1946, the first tournament after World War II, he was again one of four Americans. In 1938, when the tournament coincided with the PGA Championship—which it often did during this era—zero Americans were in England playing at Royal St. George's.

At this time the Open Championship was also not attractive

financially for players coming from the United States, when the price of cross-Atlantic travel and an entry fee only guaranteed players a spot in a one-day, 36-hole qualifier at some course near the championship venue. Even a good showing in the Open Championship was likely a financial wash, as the 1960 tournament had a total prize pool of £7,000 (about $24,000 then, the equivalent of about $200,000 in 2018), of which the winner received only £1,250 ($4,440 then, $36,000 now).

Most Americans were also unfamiliar with links golf courses, which were generally treeless and relatively close to the ocean, making for severe wind and rain as omnipresent obstacles. Scottish caddies say that during a normal round at a proper links, you will experience all four seasons. A proper links is defined by being located on the sandy soil that was least suitable for farming, between the ocean and the flatter, more fertile soil inland. Over centuries, the wind swept the sand into large dunes, held in place by wispy fescue grasses. The turf underneath was firm—perfect for bouncing golf balls along the ground—and the sand easily absorbed rain. The best use of the land was for the grazing of livestock, mostly sheep. The sheep dug out the bottom of the swales and exposed the underlying sand to use as shelter when the weather got rough. When the bored shepherds began to hit rocks around with their shillelaghs, then began to compete by seeing how many swipes it took to get the rock into a faraway rabbit hole, those sheep "bunkers" became interesting obstacles. The rocks naturally bounced to the low point of the land where the sand was exposed, and the first bit of golf course strategy was born.

"A hazard placed in the exact position where a player would naturally go is frequently the most interesting situation, as then a special effort is needed to get over or avoid it," renowned architect Alister MacKenzie would write in 1920.

These general characteristics of wind, rain, and bouncing balls were foreign to most Americans by 1960; most courses in the United States were inland, surrounded by trees, and waterlogged by soil made up mostly of clay and stone. The bulldozer had come into wide use, and the massive moving of earth was embraced as a modern concept. Most American pros were quick to complain about the "fairness" of a golf course because they were trying to scratch out a living and didn't want a bad bounce to cost them even the slightest bit of money. So they almost always decided to stay stateside during the Open Championship, and that practicality might have helped diminish one of the tenets of the early game—the joy of an unpredictable outcome.

"It is much too large a subject to go into the question of the placing of hazards," MacKenzie wrote, "but I would like to emphasize a fundamental principle. It is that no hazard is unfair wherever it is placed."

Yet the one man who mattered most to the game in 1960 felt a draw to that tournament at St. Andrews. Arnold Palmer had already begun his rise in popular culture, his blue-collar demeanor and movie-star looks making him a hero of early television, while his victories at the 1958 and 1960 Masters, as well as the epic 1960 U.S. Open at Cherry Hills just a month before—when Palmer won $14,400 for first place after edging an aging Hogan and young amateur Jack Nicklaus—had begun to build his legacy as one of the all-time great players. Palmer was never prone to an overly sentimental look at history, yet he knew something about the Open Championship was special, and something was special in particular about the Old Course at St. Andrews, ground over which the game had been played in some form for at least the previous five hundred years. (Six more courses were eventually built under the umbrella of St. Andrews Links Trust, but the Old Course remained bedrock.)

"It was exciting for me because I was trying to fulfill a desire that I had to play in the Open Championship," Palmer said in 2010 upon returning to St. Andrews. "I felt that if you were going to be a champion, you couldn't be a champion without playing in the Open and hopefully winning the Open."

Not surprisingly, Palmer couldn't conquer St. Andrews in his first try, getting hammered by a strong rain and losing to Australia's Kel Nagle by one shot. "The wind blew, it rained, and I said something about it then, and I got the same answer: 'Hey, this is Scotland and you've got to expect it,'" Palmer said. "And I loved it."

He returned next year to Royal Birkdale in England as one of six Americans in the field and won. He won again the next year at Royal Troon in Scotland—getting revenge by beating the second-place Nagle by six shots. That was also the first Open Championship that Nicklaus played in, shooting 80 in the first round and 79 in the final round to finish in a tie for thirty-fourth, 29 shots behind Palmer. Yet the interest in the tournament, and the interest in links golf, had already risen in the United States.

By the time Nicklaus won his first of three Opens in 1966 at Muirfield, seven Americans finished in the top fifteen. By 1977, the year of the famed "Duel in the Sun" between Nicklaus and fellow American Tom Watson at Turnberry, eleven of the top twelve players were Americans. By 1978, when Nicklaus avenged that defeat to Watson with his third and final Open victory, likely the most cherished because it was at St. Andrews, the field had twenty-eight Americans.

In less than two decades, Americans had opened their eyes to the wonderful history of the tournament and the beauty of links golf. This beginning of a groundswell took decades to come to fruition, as Americans gradually embraced golf courses that emphasized the natural terrain. That template always starts with the Old Course at St. Andrews, the singular golfing landscape that has withstood the test of time. It has endured less on mere historical merit and more on the substance of the holes—the enjoyment of the strategy that was not dictated by man, but by nature.

"If you're going to be a player people will remember," Nicklaus said, "you have to win the Open at St. Andrews."

One of the most intriguing aspects of golf is that it is a rare sport with vastly different playing fields, and play is even predicated on that variability. A round of golf is an ever-changing endeavor, and what works one day—or one minute—might not work the next. A talent for adaptation, along with preparation that fuels improvisation, is as crucial as any technical skill to be successful on a great golf course.

What connects golfers to courses is essentially about art, about what type of design makes a connection with the observer. And because in golf the observer interacts with the art form, the practical usefulness of the art is also integral to that connection, which can reach great depths between man and his surrounding environs. Imagine being able to throw a cocktail party in the garden of Versailles, or to go to a painting class in the Sistine Chapel.

"The attachment is from there being a sense of place," said Gil Hanse, whose celebrated career as a course architect included the distinction of building the Olympic Golf Course in Rio de Janeiro for the 2016 Summer Games, the first Olympics to include golf as an event since 1904. "There is a sense of place that just feels right. [The golf course] is in concert with its surroundings. I think there is an innate ability in all of us to know when something just feels right, feels comfortable."

Yet golf courses weren't always created in harmony with their surroundings, especially in the United States. Just after World War II, as the country eventually settled into the prosperous times of the 1950s and early 1960s, golf courses were starting to be built again. By the time Palmer went to St. Andrews, most of the top golf courses were being designed by two men: Robert Trent Jones Sr. and Dick Wilson, two designers who were coveted for being "modern." Their work was in direct contrast to those who came before them. Golf's golden age of architecture ranged from the 1910s into the 1930s, led by the founders of the art: Dr. Alister MacKenzie, C. B. Macdonald (along with his partner Seth Raynor and construction maestro, Charles Banks), A. W. Tillinghast, Donald Ross, William Flynn, and George Thomas. Back then, the architects' focus was on the land as it was found, largely because a donkey pulling a cart can only move so much dirt. With the proliferation and commercialization of heavy machinery after World War II, construction became quite a bit easier and certainly a lot more affordable.

The postwar golf courses being built were a reflection of that. Robert Trent Jones could manipulate the rolling hills in Minnesota and build Hazeltine National (1962), or dig slopes around a meandering creek to build the big greens of Bellerive in Missouri (1960), or take a dead-flat piece of land in Ohio and build some interest into a back-and-forth track such as Firestone North (1969). The same could be said for Wilson's work, a lot of which was done down in Florida, such as routing around a few lakes to build Bay Hill in Orlando (1961), or finding suitable playing grounds in a low-level area to build the Blue Monster at Doral in Miami (1962).

It's not to say that these golf courses weren't terrific, it's just that they were distinctly man-made, with features that were supposed to stand out as a testament to what the builder was able to accomplish. This design strategy often forced the player to hit all-or-nothing shots, with severe penalties for a lack of execution. Such design left few decisions for the player to make, with only one proper line of play. Games demanded perfection over imagination.

"We also have to remember that Robert Trent Jones, Dick Wilson, architects of that era, they knew Donald Ross," Hanse said. "He was a competitor. He wasn't a deity to them. He was like the guy I need to [beat out]. So now we put those [golden age] guys on a pedestal, while these guys looked at them like, 'Okay, they're old-school, and now everything is modern.' And you think about what was going on in society—automation, automobiles, postwar economy, 'Happy Days.' Everything was supposed to be newer, bigger, better. Architecture, I think, tried to embrace that. It's not to say those courses were bad, it's just they weren't quite as good."

Another architect emerging around this time was doing something a little different. Pete Dye started out designing courses in the vein of Trent Jones, but then he ran into an Alister MacKenzie course at the University of Michigan, which helped inspire a trip to Scotland in 1963. Dye fell in love with tiny pot bunkers, small and fearsome greens, and wood-boarded bulkheads to keep steep bunker faces from collapsing. That would eventually become Dye's signature characteristic—the "railroad tie" that was used wherever possible, either in a bunker, around a green, or along the water.

Dye found success with his 1964 design of Crooked Stick Golf Club in his native Indiana, a course that would eventually play host to the 1991 PGA Championship won by John Daly. After Dye had collaborated with Nicklaus on The Golf Club near Nicklaus's home in Columbus, Ohio, the two got together again for what would become one of Dye's defining designs, the Harbor Town Golf Links in Hilton Head, South Carolina, opened in 1967. The PGA Tour went there for a tournament in 1969, won by a forty-year-old Palmer, his first win in more than a year. The course was embraced for being so different. It was short but tight and had railroad-tie bunkers surrounding small greens. The four par 3s could be as good as any set of one-shotters in the world. The course was meant for precision, not power, and proved to be quite challenging. With its candy-striped lighthouse behind the eighteenth green, a hole that plays idyllically alongside the marshy Baynard Cove, the course has continued to draw some of the best players in the world every year. In 2012, PGA Tour players voted it their second-favorite course on the Tour, behind only Augusta National.

Dye would go on to build more iconic courses, including the

Stadium Course at TPC Sawgrass in Ponte Vedra Beach, Florida, the first course developed and owned by the players of the PGA Tour. If Harbor Town began Dye's legacy, then Sawgrass cemented it. It featured such a radical design and such a blatant show of man's earth-moving powers. Fifty thousand cubic yards of earth were heaved up to create the pond that abuts the sixteenth green. Water then surrounded the famous island green on the short par-3 seventeenth, an idea suggested by Dye's wife and codesigner, Alice, when they had only seventeen holes routed on the land. A green totally surrounded by water—a hazard with no recovery shot possible—is a design concept so unnatural some thought it abhorrent (and some still do). The course was long and tight, with double doglegs bending around hanging palm trees to go along with huge waste bunkers and tiny pot bunkers. The undulating small greens rolled into swales cut at fairway height, with the soil dug out to create those swales generally piled up nearby in artificial mounds. At the behest of PGA Tour commissioner Deane Beman, who pushed for the course to be the permanent home of the Players Championship, it was also the first course designed specifically with spectators in mind (as would all the subsequent courses dubbed TPC—Tournament Players Club—owned by the Tour). Therefore, the grass amphitheater that encases the seventeenth hole made for a dramatic setting never before seen in golf.

Dye was trying to get into the heads of the best golfers in the world, and he succeeded. At the first tournament held there in 1982, the complaints were aplenty. Ben Crenshaw, who missed the cut and would go on to have his own illustrious career in architecture, called it "Star Wars golf, designed by Darth Vader." Nicklaus, who also missed the cut, said, "I've never been very good at stopping a 5-iron on the hood of a car." Dye even called the players "chicken" for playing away from some hole locations.

The inaugural tournament was won by Jerry Pate, a terrific player during his time and a showman by nature who played with an orange golf ball, as if he had aptly stolen it from a miniature golf course. Walking down the eighteenth fairway with victory in hand, Pate turned to the television camera and said, "I wasn't trying to beat the field, I was trying to beat Pete Dye, and I believe I got him today. I already told him I'm putting him in this lake." After Pate tapped in his birdie, he pushed Beman and Dye into the water next to the final green, then jumped in after them. Calling the tournament for CBS, legendary commentator Vin Scully said it was "perhaps the wildest moment in the history of any professional sport."

Dye's career continued well into the 2000s, and he built more iconic golf courses with the same penchant for penalty. The 1991 Ryder Cup was held at his Ocean Course at Kiawah Island in South Carolina, and with more double doglegs and the wind howling off the beach, it proved to be one of the toughest tests of golf in the world. By then, Dye had become the icon of modern architecture, building memorable holes out of totally nondescript land. He was brilliant and brutal in his designs, meeting the era's insatiable desire for "championship" golf courses, when most owners wanted new projects to be as hard as possible.

Budding architect Tom Doak had been an intern for Dye, and in 1984, he was helping Dye work on the plans for the course at PGA West in La Quinta, California, what many consider to be the sequel to Sawgrass. The two began talking about Pine Valley in southern New Jersey, consistently ranked the number one course in the world once ranking courses became a marketable item for *Golf Digest* in 1966. Dye drove home a very poignant message about the course that Doak carried with him going forward.

"At Pine Valley, the souvenir you could buy was a set of postcards of every hole," Doak said. "Mr. Dye said, 'That's the pressure now, everyone wants that. And there's only one golf course in the world where you really get that.' That's their not-too-subtle way of reminding you that they're the one."

That mentality inspired Dye's iconic designs. He created a genre that was often replicated (to poor results) but never duplicated. Golf courses had strayed wildly from the idea of embracing nature. In the materialist age of the late twentieth century, people seemed to long for courses with pomp and circumstance rather than the courses that encouraged simple enjoyment of where the game had begun.

Dye's legacy spurred architects to take greater risks. With the mechanical ability to reshape earth in revolutionary designs, at least imagination had been revived. Soon, that mechanical power would be used to re-create nature rather than to try to surpass it.

Going back to the early days of golf course architecture, there had always been three different schools of design: the penal, the strategic, and the heroic. The schools often overlapped, but some tenets hold up in each one.

In the penal school, a single line of play is clearly outlined, and any shot hit off this line is penalized with a hazard. Trent Jones built countless holes such as this, with two bunkers pinching the fairway around the landing area on either side, and then a slightly elevated green that was surrounded by some assortment of bunkers on all sides.

"The line of play should be obvious from the tee," Trent Jones said.

A strategic design forces a player to evaluate his ability, then decide how much risk he is willing to take for a greater reward. A simple example would be a dogleg left with a deep bunker at the inside of the turn and a big mound blocking the view of the green from the right. This would make the left-hand side the better approach, meaning the closer you could hit the ball to the bunker, the better shot you would have into the green. But the width is sufficient to avoid the bunker totally, leaving a more difficult approach to the green, over the big mound. This design school forces the player to think about the next shot, not just the one in front of him.

MacKenzie first identified these two schools in the early twentieth century. His hope back then was that the once-in-vogue penal school was dying. "The trouble in those early days was that all golfers except a very small handful of pioneers belonged to the penal school," he wrote in 1933, never one to be short on recognizing his own greatness. "Today, we have no such battles to fight. I hardly come across a thinking member of a committee who does not belong to the strategic school."

Later a third school of design, known as the heroic, melded these two. Dye was a master of this type of design, where often a grand hazard defined the hole, and if you chose to take on the hazard and succeeded, you would be greatly rewarded. Imagine a short par 5 with a pond in front—hit it over the pond in two and have a great look at eagle, whereas if you come up short, par will be difficult. Pine Valley, for one, is predicated on thrilling, heroic shots.

As Doak wrote in his first book, *The Anatomy of a Golf Course* (1992), "From the point of semantics, all architects would prefer to be identified as members of the strategic school of design. Yet the best golf courses are those which borrow from all three schools." The perfect example Doak gives is the famous three-hole stretch at MacKenzie's Augusta National known as Amen Corner. The par-4 eleventh hole is a strategic wonder—even if it lost some of that strategy with the huge increase in distance added in the 2000s. As originally conceived, the hole has enough fairway to hit it left and avoid the trees right, but an approach from the left side means having to hit it directly over the small pond that fronts the green. The right side of the fairway affords the better angle into the green, severely sloped from right to left and funneling down toward the water. With a ton of undulating fairway to the right of the green, it's easy to bail out away from the water, but that leaves a terrifically difficult pitch shot (unless you're Larry Mize, who holed out a 120-foot pitch from there in a playoff to win the 1987 Masters). From the eleventh, the difficult par-3 twelfth is a full-on penal hole, requiring a shot over Rae's Creek to a shallow green, slightly angled away from the tee and surrounded by bunkers. It also sits in an area of the golf course where the winds swirl, making for one of the most harrowing 150-yard shots in tournament golf. The trifecta of holes is rounded out by the heroic par-5 thirteenth, short enough as a dogleg left that most can try to hit it in two. But the second shot is from an uneven lie in the fairway, with the creek again meandering in front of the green. But if you can hit the green, severely canted from back to front, it is receptive and can feed the ball close for an eagle putt.

This stretch of diversity in design is one of the reasons Augusta National has held up so well over the years, creating the emotional swings the players—and therefore the fans—go through during the round. Play that combines making the correct decisions (strategic), executing difficult shots (penal), and taking on risks (heroic) elevates golf to the height of its mental and physical enjoyment. Too much of the same school can be monotonous—especially the penal.

When George Crump decided to build Pine Valley in a sandy forest of New Jersey, just east of Philadelphia, he brought in many big names to consult on the design, including Harry Colt and Charles Alison, along with Tillinghast, George Thomas, and Walter Travis. The result was a difficult course with expansive natural waste areas of sand and brush. Surrounded by the dense pine forest, it was beautiful and intimidating. But it had no shortage of strategy, either, such as the short par-4 twelfth hole. The green sits to the left over a line of brutally penal sand, so anything right is going to be safe and farther away, and the closer the drive is to the sand, the closer to the green— with anything in the sand likely resulting in a bad score on what is seemingly an easy hole.

Such wonderfully thoughtful designs characterized golf course architecture's golden age. In the booming postwar real estate market, however, golf courses became an attraction, an anchor for a development. The design of many of these courses was an afterthought, like putting in a communal pool. The best land went to the most expensive houses, and the golf course likely got the worst pieces of the property. Courses were strangled by the surrounding residences, and riding in electric carts—and their subsequent concrete cart paths—became integral to the experience. Such lazy design was absolutely antithetical to the previous spirit of the game and the courses that exemplified it.

This design trend grew in concert with the rising economy. The development of golf courses and clubs became a contest of new money, with each one trying to top the next in extravagance and difficulty. From 1970 to 1989, eight golf courses were built that remained in the 2017–18 list of *Golf Digest*'s Top 100 in America. In contrast, twenty-four of the Top 100 were built in the 1920s alone—three times as many.

Throughout this boom, a minority of golfers were still in love with the classics and scoffed at the new age building. Crenshaw, for one, was always outspoken about the new courses and was detailed and thoughtful about why he disliked them. As his playing career began to wind down, he became interested in designing his own golf courses. He teamed up with established architect Bill Coore, and the two collaborated to open two new courses in 1991—Barton Creek in Crenshaw's hometown of Austin, Texas, and then the Kapalua Plantation Course on the Hawaiian island of Maui. The courses were anti-extravagance, which is not to say that they weren't breathtaking. Kapalua featured such extraordinary views of the ocean and such wonderful natural elevation changes that it seemed as if Coore and Crenshaw had simply found golf holes where they already existed. That was exactly what they wanted the course to look like and play like, too. If they shared one main design philosophy, it was that the natural land should determine the golf course.

Their big break would come after Crenshaw saw a photo of the untouched rolling sand dunes in Nebraska in a book called *The Golf Course*, by *Golf Digest* architecture editor Ron Whitten. A picture of untouched land being in a book about golf courses was strange, but Whitten was educated on the subject and knew the land was perfect for a course. Crenshaw immediately knew the same, so he reached out to Whitten, who got him in touch with the man who was scouting the land, Dick Youngscap. Deeply interested in creating an unadulterated golf experience, Youngscap hired Coore and Crenshaw, they found the exact piece of land they wanted in that area, and in 1994 they opened Sand Hills Golf Club. It was fifteen miles south of Mullen, Nebraska, which is three hundred miles west of Omaha. Mullen had far more grazing cattle than people—the town's population even in 2016 was just 481 souls. The only sounds heard amid the dunes were meadow-larks. The location came close being to the exact middle of nowhere, but Youngscap said they were trying to appeal to golf "purists," so that's the way the course was designed.

"We tried so hard to lay this golf course out on this landscape in such a quiet sort of way, without changing it any more than we had to," Coore told PBS in 2012. "We've tried to do everything here in such a way that, what was brought here to the site by us, the elements for playing golf, is indistinguishable, as much as possible, from what was already here."

The result was spectacular. By the 2017–18 rankings, the course in the middle of nowhere was No. 9 in the world, the only course ranked in the top fourteen built after 1933. Sand Hills would eventually be regarded as a huge breakthrough in golf course architecture. But really, it was a revival.

"It isn't anything new," Youngscap told *CBS Sunday Morning*, "it just hasn't been done in eighty years."

The bunkers were generally enormous, mostly made by scraping some of the topsoil off to expose the underlying sand—just like a sheep. The huge sandy expanse next to the eighteenth green had already existed, and with dots of shaggy turf interspersed, it looked just like Pine Valley. The greens sat in the most natural of places in the nooks of dunes, and the fairways rolled with the craggy irregularity of the terrain, just like at St. Andrews. The only building that can be seen from the whole property was a small wood shack behind the eighteenth green, with a deck known as Ben's Porch.

Sand Hills was the beginning of a movement called minimalism.

Historically, golf has weathered many ups and downs from the fluctuating numbers of those who play the game recreationally. During the good times, it seems players have too few golf courses to quench their insatiable thirst for the game. During the bad times, a golf course seems to be an utter waste of land and resources.

People stop playing golf for three main reasons—it takes too long, it costs too much, and it's too difficult. All of these reasons can be connected to the golf courses where most people were playing during the times that led to downturns.

When a developer wants to build a championship golf course, the new course must be long enough for Tour professionals, which requires more land. Purchasing, maintaining, and irrigating a championship-size course carries a high price tag. Moreover, as golf club technology accelerated exponentially and balls flew farther with every hit, courses needed to be even longer. That cost was put back onto the customer in the price per round, whether in a daily greens fees or a club membership. Courses tailored to the pros are also extremely difficult, which even a mid-handicapper won't enjoy playing too often. Such an inflated level of difficulty proves even more discouraging to a beginning golfer, who will think that the game is too hard to learn.

For these reasons, pros don't always make the best architects. The most notable example is arguably the best golfer of all time, Nicklaus, who collaborated with Tom Doak on the supremely beautiful Sebonack Golf Club on the east end of Long Island, which opened to rave reviews in 2006.

"Some days, I feel I'm one of the only architects who understands that a lot of people play it just for fun, as an escape, and not competitively and always trying to make their best score," said Doak, who is small in stature and wavers between a 10–15 handicap, and who had more than a few disagreements with Nicklaus during Sebonack's design. "Architects have been great players, and they think that's what golf is all about. And you can see it in their golf courses. You just have to grind all the time.

"I appreciate the walk and the view. That drives some of my design decisions."

Not coincidentally, when Doak was in college, he began trading letters with Crenshaw and meeting up with him at tournaments to watch practice rounds and talk golf. The architectural sensibilities were passed down, and when Doak applied for Cornell's postgraduate Frederick Dreer Scholarship, Crenshaw wrote him a letter of recommendation (as did Dye, Beman, and Herbert Warren Wind). When Doak started his own company, Hanse was an intern and Doak wrote the letter for Hanse's scholarship. Down the line the sensibilities spread.

The ability to articulate these ideas was also a common thread. Hanse made the comparison between the style of competitive play from someone like Crenshaw and Tom Weiskopf against that of German technician Bernhard Langer or even Nicklaus, and how that translates into design.

"There are stories I've heard about Bernhard Langer, where back in his prime—and I'm sure it's still the same—his caddie would give him a yardage, 'One fifty-eight from the sprinkler head.' And [Langer] would, in all seriousness, say, 'From the back of the sprinkler head or the front of the sprinkler head?'

"That program, that mind is thinking something completely

different than what Ben Crenshaw is thinking, who's looking around at the butterflies and the trees and then he's, 'Okay, what do I have to hit?' Feeling the shot and feeling the putt. So who are the best player-architects, and are they art or science? I think you're going to find that the best ones are artful players. Weiskopf and Crenshaw, guys that were just kind of a lot of feel and not necessarily mechanical. Where Nicklaus was more mechanical and his architecture reflects that."

When the economy was good, the demand was for big-name architects such as Nicklaus to build something impossibly pristine and lush and difficult. Yet as the economy shrank, that model became unsustainable, and it turned people off from golf entirely.

"The worth of a golf course cannot be judged on mathematical lines," MacKenzie wrote in 1931. "The crucial test is what gives the most lasting and increasing pleasure."

The design dysfunction went far outside the American borders, too. According to the *Independent* newspaper in the United Kingdom, Japan went from having a few hundred golf courses in the early 1970s to having more than two thousand courses by 1998. Not too surprisingly, the overexpansion led to the collapse of the market, and large financial firms ended up coming in and buying unsuccessful golf courses for pennies on the dollar, eventually selling them off for real estate development or other investments.

The golf boom took off even further in the United States after Tiger Woods won the Masters in 1997. Marketers and company executives bet that golf was now going to become cool to urban teenagers, and its popularity did rise for a little while. But eventually the new trend dwindled out, and it crumbled further when the entire American economy crashed in 2008.

According to a study by the nonprofit National Golf Foundation (NGF), the United States had a peak of 16,052 golf courses in 2005, which diminished to 15,372 by 2015, and to 15,014 by 2016. The country was losing 150 to 200 golf courses a year, with one NGF staffer saying, "The dirt is worth more than the grass." Yet the foundation hoped that the contraction was just a market correction for overexpansion.

"NGF views the slow and steady reduction of U.S. courses as the natural economic response to the opening of more than four thousand new golf facilities between 1986 and 2005," Greg Nathan, NGF's chief business officer, said in 2017. "This gradual reduction is indicative of the market's healthy self-balancing of supply and demand, and a trend we expect to continue for several more years."

In 2008, just over 29 million Americans had played at least one round of golf in the previous year. By 2016, that number was down to 25 million.

After 2008, Americans suddenly had far less money for recreation, especially for an expensive sport such as golf. Most of the golf market began to rethink its strategy in hopes of staying alive. Golf needed to become more affordable, as well as more enjoyable for every demographic, not just wealthy businessmen.

So minimalism began to gain momentum. Slowly, people started to embrace the idea that golf can be played on a course that is not all lush green, and that brown grass can be a terrific playing surface. The USGA, led by Mike Davis, even brought their crown jewel, the U.S. Open, to the burnt-out fescue of Chambers Bay hard against the Puget Sound in 2015. The course featured wide, bouncy fairways that made it look unlike any other U.S. Open in history. There was a goal behind that.

"Through research and best management practices, we see a future where golf courses can reduce their consumption of critical resources, such as water, by twenty-five percent by 2025," Davis said as the CEO in his 2018 keynote address to the USGA. "This will not only add to the protection of water resources, but better financial bottom lines for golf courses. Here in the U.S., the golf community already realized a twenty-two percent drop in irrigated water usage, and a forty percent drop in nutrients over a recent ten-year period, thanks to best practices and research advanced by the USGA and implemented by countless golf course superintendents across the country."

But golf courses must be purposefully designed to use less water. Courses are frequently built with bent-grass greens, for example, in the tradition of so many historic golf courses in the Northeast. "If you don't water bent-grass greens, bye-bye, they're gone," Davis said. But if a course is consciously designed in the name of sustainability, then cost savings translate to the consumer. Courses that require less maintenance also offer a more enjoyable round for the average player, having fewer hazards and less high rough so close to the playing lines.

"The real objective of golf architecture is to make the course difficult for the better golfer without making it too difficult for the hacker. In USGA parlance, this would be a course with a high Course Rating but a low Slope Rating—a very rare combination to find," Doak wrote in an essay titled "The Best Hazards in Golf: Short Grass as a Golf Hazard." The USGA established its twofold rating system by 1981, using *course rating* as a number for how the average scratch golfer (zero handicap) would score, and the *slope rating* for how a bogey golfer (18 handicap) would score.

"As such," Doak wrote, "the ideal hazard would be one which scares the knowing, but of which the average golfer is not aware."

For a long time, the demand for this type of less elaborate course was almost nonexistent. But once sustainability became an integral piece of the economic picture, this type became far more attractive. The ethos behind original links golf courses was again resurgent, and Sand Hills inspired a multitude of contemporaries.

In 1999, Mike Keiser, who earned his fortune making greeting cards, opened the first course at the Bandon Dunes resort on the Oregon coast. It was five hours south of Portland, and the closest airport was forty-five minutes away, a one-runway outfit in a town called North Bend. He hired a Scotsman, David McLay Kidd, to build the resort's namesake track, and it was spectacular. A large portion of the land ran along a rugged coastline that dropped down a hundred-foot cliff and opened to a two-hundred-foot beach before reaching the crashing waves of the Pacific Ocean. Players standing on a tee box could see migrating whales swimming just past the breakers. The ground was sandy, with yellow gorse bushes growing in the hills, just like on the links of St. Andrews. It seemed like a modern version of a century-old links course in Scotland or Ireland.

As part of his grand idea, Keiser had also bought a lot of the surrounding land, and he quickly developed it. He brought in Doak to build Pacific Dunes, which opened in 2001 and has continually stayed inside the top twenty on the *Golf Digest* list. He brought in Coore and Crenshaw to develop the inland part of the property, and Bandon Trails opened in 2005. Coore and Crenshaw also designed a thirteen-hole par-3 course over the most rugged piece of land, calling it the Preserve—with all proceeds going to the Wild Rivers Coast Alliance, dedicated to preserving the environment of the southern Oregon coast. Doak and historian Jim Urbina also designed a course called Old Macdonald, a tribute to the work of C. B. Macdonald.

Not by coincidence did Doak name his company Renaissance Golf Design, and this embodied the rising tide of sentiment that brought the game back to its roots. Keiser opened another destination golf resort in Wisconsin called Sand Valley, and Doak designed a course at a similarly spectacular "purist" destination, this one called Streamson in central Florida that also housed a course designed by Coore and Crenshaw, and another by Hanse. The ethos was alive and well.

Whether out of financial necessity or nostalgia, people once again seemed to care about the golden age. One of its leaders, Macdonald, based his work on holes he played in Europe while learning the nuances of the game. He brought those designs back to America and recreated them all over his courses, imaginatively using these "template" designs as bedrock for what would become his loved classics.

There was the Redan, a par 3 with a deep bunker (or bunkers) short left and a green that is severely canted from front right to back left, normally with a fairway run-up. It was based off a hole at North Berwick, in Scotland. Another was the Biarritz, a long par 3 with a green that has a giant swale in the middle. It was based off a hole designed by Willie Dunn in Biarritz, France. Another would be the Alps hole, a blind par 4 based off a hole at Prestwick, in Scotland, and another would be the Leven, a short par 4 based on a hole at the old Leven Golf Club in Scotland (now called Lundin Links).

These holes defined the strategic school of architecture, the holes that stood the test of time. They traced their roots back to the most natural landscapes in golf, and their connection to nature and history is what drew the interest of many. Most of the work commissioned of the high-end architects in the 2010s were "restorations" of golden age courses. Among others, Doak restored two MacKenzie classics in California, Pasatiempo and The Valley Club. Hanse did all the redesign work at Tillinghast's 36-hole masterpiece in New York, Winged Foot. Coore and Crenshaw did the massive renovation at arguably the best golf course in America, Flynn's Shinnecock Hills, which then held the 2018 U.S. Open.

Such restoration work was a necessity for these design firms to stay alive. In 2016, only ten new golf facilities were built in the United States. But restorations and renovations of old courses still fueled the fire of those who longed for a more intelligent design, one that echoed history and embraced nature.

And so it goes back to the originator of the game, St. Andrews, where Macdonald pulled two of his template designs from the Old Course. The first is the Road Hole, a long dogleg par 4 with a narrow, diagonal green split in front by a brutal pot bunker and backed by a hazard. The hazard on the seventeenth hole at St. Andrews is a public road, with an ancient stone wall behind it. The other is the Eden, a par 3 with a green sloped hard from back left to front right and with one bunker in front and one on the left side of the putting surface. In that left bunker on the eleventh hole at St. Andrews—known as Hill bunker—a young Bobby Jones once took a handful of swipes, then ripped up his scorecard and picked up his ball, quitting the 1921 Open Championship. Jones called this moment "the most inglorious failure of my golfing life."

Of course, Jones went on to break through at the 1923 U.S. Open at Inwood on the southwest part of Long Island and eventually won thirteen major championships (this when the U.S. Amateur and British Amateur were majors). He completed the Grand Slam in the single season of 1930, then quit competitive golf altogether. That British Amateur was at St. Andrews, and his time in that place stuck with him for the rest of his life—including while he built Augusta National with MacKenzie.

"I could take out of my life everything except my experiences at St. Andrews," Jones said, "and I would still have a rich, full life."


What makes a golf course great, makes the player connect with the land and the design so deeply? Why is there an obsession with great old golf courses, and why has that attachment to history regained such popularity?

The singularity of golf course architecture holds the answer. Being at St. Andrews is not just like being at a golf course. It is more than that, and it is less than that. It is feeling nature and man in communion. It is fun and difficulty in unison. It is reveling in how the people of the town have kept their Holy Grail intact for all these centuries, and listening to the audible ghosts in the grass.

MacKenzie once spent a year drawing a detailed map of the golf course the size of a dining room table. Tillinghast once played a round there with Old Tom Morris and wrote that it made him "feel as though my own game must seem glaringly new, just like walking up the church aisle in new, squeaky boots, but this feeling soon vanished. The old man and I were just boys together, for such is golf." Donald Ross grew up north of St. Andrews in Dornoch, Scotland, but felt the need to serve as an apprentice under Old Tom at St. Andrews before coming to America in 1899. Doak spent the summer of 1982 caddying at St. Andrews and exploring the golf courses of the British Isles while on the Dreer Scholarship. Hanse did the same five years later.

Since the game began, the Old Course at St. Andrews has been the center of the golfing world, for reasons both intellectually logical and spiritually coherent.

"St. Andrews," MacKenzie wrote, "is a living example of the possibility of obtaining finality."

# EIGHT

# WILLIAM JAMES AND THE BIRTH OF SPORTS PSYCHOLOGY

Bob Rotella believes that all great athletes are great storytellers.
This doesn't mean that great athletes are all articulate, but only that they share an inherent ability to deliver a message to themselves, clearly and with conviction. It also means that they have a useful imagination, conceiving the extraordinary and making it tangible. To be able to convince oneself that an unclear narrative is true—that an opponent has slighted you, that your performance was better than it was, or that the loftiest goals are obtainable—can be useful in sports. It can bring motivation, confidence, and focus. That ability to tell yourself a good enough story, true or false or somewhere in between, so that you believe it, stands as a testament to the power of mind over body.

One of the biggest things that Rotella does as the top doctor in the field of sports psychology—a field he had a huge part in creating—is to remind athletes of the stories they should tell themselves, rather than the stories they should forget. With a masculine demeanor and a penchant for simple language that belies his academic background, Rotella seeks to be viewed as a type of coach rather than what he actually is—a highly trained psychoanalyst.

"A regular psychoanalyst or a psychiatrist would take people at an abnormally low level of functioning and make them normal," Rotella said. "My job is to take someone already above normal functioning and take them to greatness. I'm teaching people how we want them to think to be great."

Rotella brought this approach with him to Florida in March of 1984, when he was invited to Miami for the Doral Open to talk to a few professional golfers who were in need of some help. One of those golfers was Tom Kite, who had at that time won five events on the PGA Tour. Although he had already cashed a lot of big checks, Kite seemed to be underachieving relative to the talent he had displayed as a terrific young golfer out of Texas. He was in a mental funk that had penetrated his golf game.

"I was in one of those places where I just couldn't seem to do anything right on the course," Kite would later write.

Rotella and Kite met a couple times early in the week, and as Kite described it, "Doc did no more than refresh my memory of those great thoughts I usually have when I am playing my best." Rotella reinforced in Kite the idea that the human will to believe is as strong as any other mental capacity, either conscious or subconscious. Thinking that good things will happen is a conscious decision. And doing so helps—a lot.

So the two reminisced about the chip Kite holed to win the 1982 Bay Hill Invitational in a playoff over Jack Nicklaus and Denis Watson. Kite hit that chip with the pin out, no fear of hitting a bad shot or possibly leaving a long putt to tie. The two pulled up the memory of the third-round 62 that led to Kite's win at the 1983 Crosby Pro-Am at Pebble Beach. Rotella asked Kite what he felt like during those moments of high performance, and suddenly Kite was back to consciously thinking about his mentality when he had played his best. Rotella made him realize that he was so positive during those times, without focusing on the overall outcome or other outside influences. It was just one shot to the next, without any set of external expectations. Rotella was helping Kite remember how his mind worked when he was most successful, and Kite started to feel it again. At Doral, under his signature wide-brim straw hat, Kite shot a final-round 65 and edged none other than Nicklaus for a two-shot victory. "My swing hadn't changed at all in the couple days since the last event," Kite wrote, "but I felt like a new person."

In 1992, Kite won his lone major at the U.S. Open at Pebble Beach, and for a long time he held the mantle as the leading money-winner in PGA Tour history. He spoke to Rotella often, and he wrote the introduction to Rotella's wildly popular 1995 book, *Golf Is Not a Game of Perfect*.

"What Doc can do is show you what thoughts are advantageous and what thoughts are destructive," Kite wrote. "And one of the really neat things that comes along when you try this approach is that not only do you become a better golfer, athlete, or sales executive, but you learn more about yourself and become a more fulfilled person.

"Who says we can't have it all?"


In the first chapter of *Golf Is Not a Game of Perfect*, Rotella wastes no time in making his most glaring declaration:

"I have two things in common with Sigmund Freud. I have a couch in my consulting room. And I ask people to tell me about their dreams. But there the resemblance ends."

Freud introduced the science of psychology with his 1899 publication of *The Interpretation of Dreams*, where he explains his concept of how the subconscious mind works and how the conscious mind filters it, and how our subconscious can manifest in our dreams. For Freud, the unconscious mind—what he would later define as the id—was driven by "infantile wishes, which are erotic and aggressive in nature," according to Daniel T. O'Hara and Gina Masucci Mac-Kenzie, two academics who wrote the introduction to the 2005 edition of *The Interpretation of Dreams*. The conscious mind, consisting of the ego and the superego, would filter those wishes into symbolism in a dream, providing a safe mental outlet for culturally unacceptable ideas such as the Oedipus complex—a childhood desire that could linger throughout adulthood, where one wants to have sex with one parent and kill the other.

Freud created a way to try to understand the symbols in a dream, and he led the dreamer in "free association," in which he mentioned a symbol and the dreamer said the first thing that came to his or her mind. Freud thought he could determine the meaning of any ambiguous symbolism and thoroughly analyze the purpose of the dream. By listening to patients describe their dreams and leading them in free association, Freud thought he had found a way to map the subconscious by understanding the messages it was trying to get through. In turn, he could better understand why humans acted a certain way.

"The dream, rather than perception or cognition, is the characteristic production of psychic reality," wrote O'Hara and MacKenzie.

Most of Freud's patients were dealing with neuroses. He was trying to help and understand people who were not only personally suffering, but who were struggling to fit into a rigid social structure. In his work, Freud used his patients' experience but also cited his own dreams for reference—an airing of his mental makeup that he found difficult, but necessary.

If that can be considered a rough outline of Freud, and, in turn, the beginning of psychology, then it is not hard to understand why so many practical-minded people thought of psychology as too strange to be useful. Many people found it to be applicably incorrect and a self-sustaining and borderline pointless segment of academia. As Rotella went through his education, he became one of those people.

"Unfortunately, major branches of psychology and psychiatry during [the twentieth] century have helped promote the notion that we are all in some sense victims—victims of insensitive parents, victims of poverty, victims of abuse, victims of implacable genes," Rotella wrote. "Our state of mind, therefore, is someone else's responsibility. This kind of psychology is very appealing to many academics. It gives them endless opportunities to pretend they know what makes an individual miserable and unsuccessful. It appeals as well to a lot of unhappy people. It gives them an excuse for their misery. It permits them to evade the responsibility for their own lives."

Rotella did align himself with one academic: nineteenth-century American philosopher/psychologist William James. Even though James had a prestigious position as a professor at Harvard University and was the first professor to offer a class in psychology, most of his ideas didn't immediately take hold. What he taught would eventually be called pragmatism, but it was often overshadowed by Freud's work, as it is now. Because Freud had begun a new field of science, many thought his work was not philosophically subjective theories, but objective observations of how the mind worked. It was a lot like TrackMan seemingly explaining golf.

At the turn of the century, James was invited to give an address to the American Psychological Association. He described the findings of the first fifty years of research into the mind in one sentence:

"People by and large become what they think of themselves," he said, then left.

James wrote a defining essay titled "The Will to Believe," in which his genius of deduction was in full force. He disputed the attack on faith, as academics said that believing a fact without sufficient empirical evidence was considered "the lowest kind of immorality." As

James argued, the idea that "truth" even existed undercut all logic and was, in itself, a matter of faith.

"Our belief in truth itself, for instance, that there is a truth, and that our minds and it are made for each other—what is it but a passionate affirmation of desire, in which our social system backs up?" James said. "We want to have a truth; we want to believe that our experiments and studies and discussions must put us in a continually better and better position towards it; and on this line we agree to fight out our thinking lives. But if a pyrrhonistic sceptic asks us *how we know* all this, can our logic find a reply? No!"

James concluded that our belief in a fact can bring that fact to life. On the surface, this claim seemed intrinsically illogical. But James helped explain his point with a simple analogy—the cohesion of a sports team.

"Wherever a desired result is achieved by the co-operation of many independent persons, its existence as a fact is a pure consequence of the precursive faith in one another of those immediately concerned," he said. "A government, an army, a commercial system, a ship, a college, an athletic team, all exists on this condition, without which not only is nothing achieved, but nothing is even attempted. . . . There are, then, cases where a fact cannot come at all unless a preliminary faith exists in its coming."

It's easy to see why Rotella was drawn to this idea of cooperative faith, because he came to psychology through sports. Growing up in rural Rutland, Vermont, the seed of Rotella's life work was planted when his cousin Sal Somma would visit from his home on Staten Island. Somma was a local celebrity of sorts, the head football coach at New Dorp High School, where he would win or share six city titles from 1948 to 1976. (In 2010, the corner of Hylan Boulevard and New Dorp Lane was named after him.)

Somma was also good friends with Vince Lombardi, who coached at St. Cecilia in nearby Englewood, New Jersey. Lombardi would go on to lead the Green Bay Packers to five NFL Championships in the 1960s, but at the time he was just a high school coach trying to support his family. Born in Brooklyn and educated by way of a football scholarship at Fordham University in the Bronx, Lombardi took the job at St. Cecilia in 1939 for a salary under $1,000 a year. He taught Latin, chemistry, and physics and eventually got the head coaching job when his buddy Andy Palau left to go back to Fordham in 1942.

For the next five years, Lombardi was the head coach at St. Cecilia, his team being recognized as one of the best in the nation in 1943, beating the powerhouse Brooklyn Prep team led by a defensive back named Joe Paterno. During this time, Lombardi would travel with Somma as the two supplemented their income by giving coaching clinics throughout the Northeast. Lombardi homed in on what would become a masterful form of motivation and, unbeknownst to him, developed a Jamesian view on the role of attitude and how it affects the outcome of games. Lombardi's ideas about attitude stuck with Somma deeply, and when he would go up to visit his family in Vermont, Lombardi was all he would talk about.

"I'd be the only one in the family who really was interested in hearing my cousin's stories," Rotella said. "I was just mesmerized by him."

Rotella grew up close to the local Catholic high school and occasionally enjoyed taking a shag bag of golf balls and smacking them around the field. But he was more focused on other sports and played basketball and lacrosse for a small school nearby, Castleton State College. He then went to the University of Connecticut, where he took a job as an assistant coach with the lacrosse team and as the head coach of a high school basketball team, all while getting his master's and doctorate in psychology. Rotella eventually landed at the University of Virginia, where he started a sports psychology program and worked closely with the Division I programs.

Although golf was interesting, his primary focus was on basketball, and sometime around 1978 or 1979, he gave a talk in New York City to some basketball coaches about the mental approach to the game. In the audience was a man who worked for Time-Life, which had owned *Golf Digest*. He was so enamored of Rotella's ideas and saw the connection to golf so clearly that he invited Rotella to speak to the magazine's "advisory board," a group of people who carried enormous clout in the game. Decades later, Rotella could still tick off their names with ease.

"It was Sam Snead, Dr. Cary Middlecoff, Paul Runyan, Bob Toski, Jim Flick, Davis Love, and Peter Kostis," Rotella said without hesitation.

The plan was for Rotella to speak to the board for an hour and a half on "the role of the mind on golf," followed by a long question-and-answer. He laughed later at how daunting that task sounded. He laughed even harder while remembering the numerous phone calls he received in the days leading up to his talk, warning him in particular about one member of the board.

"Sam is seventysomething years old, very old-school," Rotella was told. "He'll probably rip you to shreds. Don't take it personally."

Rotella flew down to the Contemporary Hotel in Disney World and walked into a conference room filled with all these great golfing minds. He gave his speech, talking about how to strive for excellence, stay confident, and think positive. He spoke glowingly about the power of free will. Everyone seemed engaged.

At the end Snead stood up—and you can almost hear the oxygen being sucked out of the room. "Boy," he said, "I'd hate to think how many U.S. Opens I would've won if I had you to talk to when I was coming up."

Everyone cracked up, and Snead's verdict was the turning point of Rotella's career. If Snead had said anything else, if he had derided the idea of mental coaches, as so many were expecting, the rest of the room would probably have followed suit. From there, sports psychology in golf would likely have been delayed for years, if not indefinitely. The stigma of weakness attached to needing a psychologist of any sort would have stuck around a little longer, and the careers—and lives—of so many ultratalented golfers would have gone unfulfilled.

Instead, Snead began to "spill his guts," as Rotella described it. Having come out of the rural Virginia hills with a swing tempo that still draws envy, Snead won the first tournament he played on the Tour in 1936. In 1937, when he was twenty-five years old, he played in his first U.S. Open, at Oakland Hills outside Detroit. Knowing he had a chance to win, Snead nonetheless convinced himself that he was trailing and ended up three-putting three times. It resulted in a 1-under 71, and Snead finished two strokes behind Ralph Guldahl, who won the first of two consecutive national championships following a spell where he had quit the game out of frustration to become a carpenter.

Snead explained to Rotella and the group how that loss stuck with him for the rest of his career. He always thought it was his best chance to win a U.S. Open, and he blew it. Although he finished second three more times, he always thought he wasn't destined to win that tournament—and he never did. Snead compiled a then record eighty-two wins on the Tour and seven major championships. But the lack of a U.S. Open, and the scars from that tournament in 1937, stuck with him for his whole career.

"And he went on and opened up about a whole lot of stuff for about fifteen or twenty minutes or so, and that led everybody else to open up and talk about it," Rotella said. "I've always wondered if Sam Snead had said, 'This is a bunch of garbage'—and he was so famous and such a great player, even with Middlecoff and Runyan there, and Toski—I always wondered if anybody else would've even wanted to talk openly enough.

"But after that, everyone started asking great questions."

Rotella went on to give talks at golf schools run by *Golf Digest*, sometimes speaking directly to the VIP students and sometimes just to the staff. He was once talking about how rookies on the Tour should believe they can win, and that raised the interest of the always-curious Runyan.

"When I went on Tour, we were always taught to really respect veteran players and to not even think of winning for five or seven years," said Runyan, whose career through the 1930s and 1940s included twenty-nine Tour wins with two majors. "We should just respect the older players and learn from them. That doesn't seem to fit with what you're talking about."

"How many times did you win in your first five years on Tour?" Rotella asked.

"Well, none."

"That's why we don't teach that anymore."

From there, Runyan asked Rotella if he could buy him dinner that night and continued picking Rotella's brain over a steak. That initial talk, and the backing of Snead, made such an impact on Runyan that he continued to pepper Rotella with questions for years to come, as would Toski and Love. Kostis liked the message so much that he invited Rotella to Miami in 1984 to speak with a couple of students he was teaching. One of them was Tom Kite.

Rotella continued to work with high-level sports teams, such as John Calipari's basketball teams from UMass to Memphis to Kentucky, as well as to lead seminars for the management teams at Fortune 100 companies. In 1984 he was the main figure in a segment on *60 Minutes* where he helped the Virginia men's basketball team, taking them through group visualization exercises on their game plan and assigning them to write papers on such things as the importance of the bench player. Golfer Denis Watson saw the segment down in South Africa, and he asked his swing instructor, David Leadbetter, to call

Rotella and set up a meeting. After meeting with Rotella for a few days, Watson went on to win three times on the Tour that year, going from the fringes of the money list to No. 4.

Clearly, Rotella's future lay in golf. He was the modern-day pragmatist, and he continued preaching his own take on William James.

Day in and day out, his approach was simple—make people believe that they can succeed. When players called him to say they had nervous butterflies, his first reaction was always "Great!" As he told *Golf Digest*, "The only time butterflies become harmful is when we let our fear of them control us. . . . Learn to love the butterflies, or at least to handle them. One way some players handle them is to downplay the importance of today's round or tournament and think of all the reasons it doesn't matter. Taking deep, slow breaths can be helpful. Visualizing what you want can be helpful. The calmer and clearer you can keep your mind, the more you can keep it focused on what you want, the more the butterflies will fade and fly in formation."

Rotella helped golfers such as Irishman Pádraig Harrington, one of Rotella's most successful students, go through these visualization exercises. Having always been an overthinker, constantly tinkering with his swing, equipment, and mental approach, Harrington was a huge beneficiary of Rotella's advice. The visualization exercise was extremely helpful when Harrington finally broke through with his first of three major victories at the 2007 Open Championship at Carnoustie. The night before the final round, with Rotella staying in the same house, Harrington went into his bedroom and visualized what it would look and feel like to win. The next day, just before he went into a playoff with Sergio Garcia, Harrington was on the putting green and told Rotella that in the coming four-hole playoff, when it looked as if Harrington would be waving to the crowd, he was actually going to be visualizing holding the Claret Jug.

"I have never heard him say that," Rotella remembered. "I said, 'Wow—he is ready.'"

This wasn't overly complicated information, at least not on the surface. Rotella had identified a philosophy—or a type of psychology—that had always worked for situations such as high-pressure performance. It wasn't new, but it was new in terms of application. He had hit something concrete in the way that great athletes think, which meant that he could teach it to those who didn't intrinsically have it themselves. It had become a better tool than any swing tip. By 2015, Rotella had counted seventy-four major championships for his students, including men, women, juniors, and seniors.

Because of the success that Rotella had, an entire industry of sports psychologists was born. Many of them found a home in golf, where so much of the difference between success and failure at the highest level is due to mentality.

Dr. Gio Valiante was an immediate disciple of Rotella's and gained notoriety with a book titled *Fearless Golf*. In it he describes scientific information about how nervousness or fear can cause a physiological reaction in the body, releasing hormones that lead to muscle tightness and restrain the body's ability to swing freely. Pia Nilsson, who was an accomplished amateur in her native Sweden, joined up with Dr. Lynn Marriott to create a company called Vision54, where they do similar work focused on mental commitment. They have worked with many high-profile students, most notably Hall of Famer Annika Sorenstam. Dr. Joseph Parent wrote a popular book called *Zen Golf*, which leans toward Eastern philosophy and focuses on "being present" in the golf game.

Even Deepak Chopra, the world-renowned self-help author and speaker, dabbled in the world of golf psychology with his *Golf for Enlightenment*. A licensed physician who was once the chief of staff at New England Memorial Hospital, Chopra became deeply entrenched in Transcendental Meditation and is a leader in the field of alternative medicine. His book tells a fictional parable of a golfer who finds that letting go of mental constraints and bitterness toward imperfection is the way to, well, golfing enlightenment.

The popularity of these books among average golfers, coupled with the proliferation of mental coaches at the highest levels of the game, showed that most people who regularly play golf intrinsically understand how much impact mentality can have on success. That idea is nothing new. To some older coaches and players, the whole discussion around mentality only adds another layer to make a rather simple game overly complicated.

"The problem is, there are too many choices now," Lee Trevino said. "We never saw our swings on video and we never talked to anybody about it because there was no one to talk to!"

Because Rotella's approach is so practical, some golfers believe that his work is unnecessary. Swing coaches, especially at the highest level, will often say that they act as psychologists more than technical coaches. With the players holding such a high degree of mechanical mastery, there is only so much one can advise.

"It's ego management," said one coach on a major-championship driving range, not wanting his players to hear that.

The work being done by people such as Rotella generally happens behind the scenes. Their coaching lays a foundation for how to think through the execution of a tournament week. While physical capabilities are the largest factor separating athletes in other sports, the ability to think clearly and strategically has been recognized as the strongest attribute a golfer can have. The science of sports psychology, following the path set forth by William James more than a century earlier, has advanced to understand how to manipulate a person's thoughts to achieve the highest probability of success.

"Though I teach psychology, I have never known for sure where the mind ends and where the heart, soul, courage and the human spirit begin," Rotella wrote. "But I do know that it is somewhere in this nexus of mind and spirit, which we call free will, that all great champions find the strength to dream their destinies and to honor their commitments to excellence."

Although it might pain James to know this, his philosophy was eventually backed by clinical research, not just taken on faith. Neuroscience has shown that belief can manifest as chemical reactions in the brain, opening neural pathways that lead to better communication and increasingly proficient motor functions. The Freudian idea that the subconscious rather than the conscious mind is the psychic reality has been at least partially debunked.

"Freud advocated a very complicated theory of how unconscious processing works," said Dr. Greg Miller, the department chair of psychology at UCLA. "Psychological science has found the richness of that theory too difficult to study fully. It's thoroughly out of favor in academic psychology and largely out of favor in clinical practice. It's now widely believed that the weight of the clinical research literature indicates that less rich and less complicated views of unconscious processing are much more tractable—much easier to study, validate, and rely on in applied contexts, such as psychotherapy, human factors engineering (e.g., presenting data to pilots, surgeons, drivers, etc.), and, surely, sports."

Just as James thought, and as Bob Rotella taught, faith in oneself and the willingness to believe can shape one's reality.

In golf, what so many have known before is being proven—that the mind can be the biggest detriment or the biggest weapon a player has. Figuring out how to use it, and figuring out how we learn, has become the most important scientific field in the game.

# NINE

# ON LEARNING: THE ESSENTIAL DISCONNECT

This is an unpopular opinion among his peers, but Mike Hebron repeats it without hesitation:

"Teaching," he said, "is actually very harmful. It exists to make money and fill magazines. Most of it is not actually helpful to the student."

This comes from a man who is a PGA Master Professional, who was National PGA Teacher of the Year in 1991, who was inducted into the PGA of America Hall of Fame in 2013, and who has worked with players who have won every major championship. Yet teaching, to him, has assumed an ugly connotation that means fixing. Fixing is an antithetical idea to the nature of learning, something Hebron—who calls himself a "coach"—realized during a "paradigm shift" in how he thought about instruction.

As neuroscience went through major leaps in understanding during what is commonly referred to as "the Decade of the Brain" in the

1990s, Hebron studied the process of how humans learn. He spent extensive time taking classes and hearing lectures at such places as Harvard and UCLA and found academic studies that proved that teaching with explicit and specific information was not "brain compatible." Relaying conscious thoughts was incredibly inefficient in creating long-term and impactful memories.

"He has a lot more evidence of what he's doing than what most of us do," Sean Foley said of Hebron, the man who Foley feels influenced his teaching more than any other.

Essentially, Hebron found that telling someone a piece of information without attaching it to something the person already knows—such as a metaphor or a story—wasn't effective in teaching, as students failed to remember or execute the abstract information. With complex motor patterns such as the golf swing, giving students explicit information was even worse at accomplishing change.

Discrediting traditional instruction meant undercutting the entire industry that made Hebron successful. What is a golf lesson if it's not the instructor telling the student what to do? Most glaringly, these studies undercut the entire process of education and how it operated for centuries. We now had a scientific explanation for why so many young students struggle with rote content knowledge in school—it's because our brains do not work that way.

What studying the brain has shown is that self-discovery is the way to reach impactful learning. The best thing a teacher can do is create a safe, playful environment where curiosity is encouraged, where *inconsistency* and *unwanted outcomes*—the preferred terms for *failures*—are embraced, and where students' motivation to figure things out on their own is cultivated. Teaching has been proven to be most effective when the focus is less on the information and more on how it is presented.

That is especially true since so much of the world's information is now so easily accessible on the Internet. Nothing being taught in schools—or on driving ranges—can't also be found online, often for free. The role of the teacher, then, is to make information stick. So rather than thinking that the goal of learning is the amalgamation of external information, it turns out that the simpler answer is more scientifically sound—human beings were built to learn on their own.

"One hundred thousand years ago, at least six different *Homo* species inhabited the earth. Today, there is just one, us, *Homo sapiens*; how did our species succeed?" Yuval Noah Harari wrote in his book *Sapiens: A Brief History of Humankind*. "Human beings are a brain-based species, wired for survival and the ability to learn, not fail."

Or, as Hebron says repeatedly, "The brain doesn't do things badly, it does bad things perfectly. We're not wired to fail tests, miss putts, or have a car accident. Fifty species a day become extinct on Earth and we're still here, because we adapt. Learning is a survival skill."

Hebron drives home just how unnecessary golf instructors are in many interesting ways. One time, he asked a waitress if she had ever played golf. She said no. He picked up a steak knife and asked her how she would hold it if she was to play. She held it in a ten-finger baseball grip. He asked what she would do if she wanted both hands to work together. She then held it in a perfect Vardon overlap grip.

"See?" he said.

By the mid-2000s, Hebron taught almost exclusively at his home course, a public place called Smithtown Landing, in northern Suffolk County on Long Island. He had frequently been offered jobs at nicer private clubs, jobs that would have come with more recognition, money, and resources. But other jobs would have also come with membership demands he might not have liked. Instead, he spent more than four decades teaching at a course just five minutes from his house, where he could set his own agenda.

So instead of glad-handing pro golfers and loitering on PGA Tour driving ranges, Hebron was absorbed in giving a lesson to a high school boy off an old plastic mat. The range went up a hill of mostly sand and scrub. To the left were bushes that barely obscured a view down to the muddy Nissequogue River. To the right were thirty children, between the ages of eight and twelve, taking part in the week-long camps that run throughout the summer and bring more than seven hundred kids through the course in three months.

The high schooler was a pretty good athlete who played on his golf team in the spring and shot around the mid-80s. He had taken lessons from Hebron before and was comfortable in the environment. Clearly he had a lot of active leg movement during his swing, took it back above plane and came in rather steep, with little wrist hinge or lag. He also tended to make a quick transition and was losing shots thin out to the right or heavy to the left. A normal instructor might tell him to squat a bit more on his back leg, "cover" the ball with his chest, and stay behind at impact. That would probably get the hands out in front of the clubhead and create more solid, consistent contact.

Hebron didn't mention any of this. Instead, he asked the student how many pieces the golf swing has, and the student kind of shrugged.

"One," Hebron said. "There is only a beginning and an end, and there is no middle. The golf swing is a one."

The student took a quick swipe and hit one low and right.

"Was that a one, or a two?"

"Two," the student said.

He made a similar swing, and Hebron said nothing. Then the student took a smooth swing and hit one high and straight.

"Better or different?" Hebron said. "This is huge. Better is an emotional proclamation. Different is just a fact. Now, come with me."

Hebron led the student twenty yards away, over some grassy dirt to a green-shingle shack. On the outside was a white sign with a red-penciled picture of a brain, captioned NEURO-LEARNING GOLF. BRAIN COMPATIBLE IMPROVEMENT. Under that was a yellow sign that read

MICHAEL HEBRON'S HALL OF FAME GOLF CAMP. Inside, the shack had a green plastic mat and mirrors on the walls. Hebron asked the student to set up and take a swing while looking at the mirror straight ahead—a side view of his swing. Hebron asked him if he liked what his legs were doing.

"No," the student said.

"Why?"

"I think they shouldn't move as much."

"Okay."

The two stormed out of the shack.

"That's the impact of the mirrors as compared to video," Hebron said as an aside. "It's self-learning."

Back out on the range, a Ping-Pong table was surrounded by balls strewn all over the ground. It was used (mostly) for the children at camps. "You want to learn how to control the face?" Hebron said, mimicking a swing with a paddle. To the right was a disconnected circle of hard rubber blocks. Children were walking on them like balance beams, then jumping from one segment to the next—as if the ground were hot lava. "Balance," Hebron said. "You teach the athlete first." Up the hill, circling around the eighteenth green, the first tee, and the humble pro shop, was a practice putting green. In the rough, Hebron set up the popular beanbag game cornhole. "For students struggling to chip or putt, just do this." He tossed two bags near the target hole and then the third right in. "Self-discovery," he said. Hebron also had a basket full of tennis balls that were attached to three-foot-long blue-and-orange lanyards. An old friend and teaching pro had designed the device, and Hebron bought as many as the friend had made. Holding the lanyards, Hebron slung the tennis ball underhand toward a hole on the green. Then he picked up a wedge and hit a high flop shot with a similar motion and tempo. A man was on the putting green missing the same three-foot putt over and over. "That guy," Hebron said, "would be much better served doing this."

This is what unconscious learning looks like. It is *implicit* instruction rather than *explicit* instruction. It is fun and playful. It allows students to understand that they intrinsically know how to execute the motions that create good golf shots. "People know what balance, timing, and rhythm are," Hebron said. What he was doing was accessing the part of the brain that is far more powerful.

This type of methodology has existed in the fringes of golf for a long time, although people (including the PGA of America) have dismissed it as too simplistic. Hebron's pedagogy can trace its roots back to Ernest Jones, an Englishman who lost his right leg in World War I but went on to a terrific career in golf instruction in the United States, teaching the likes of Lawson Little and Horton Smith, with the message carried down through Hall of Famer Manuel de la Torre. Jones's main focus was to "swing the clubhead," and he taught the concept with a penknife at the end of a handkerchief. The goal was to think only about that general result and let the body take care of the details. "The essential thing for the mind to dwell upon," Jones wrote in 1920, "is not what movements take place, but how and where to apply power. For if power is properly applied, the accessory or accommodating movements are not likely to give trouble."

In Jones's time, this was a postulate. Decades later, Hebron could cite scientific studies to back it up, most notably in his 2017 book, *Learning with the Brain in Mind: Mindsets Before Skillsets*. The book was coauthored by Dr. Stephen Yazulla, a professor emeritus in the departments of neurobiology and behavior, as well as ophthalmology, at Stony Brook University, on Long Island.

"The non-conscious capacity of people to acquire information is much more sophisticated and rapid than their conscious capacity to do this," Dr. Joseph Weiss wrote in 1997—which Hebron and Yazulla cited in 2017—shortly after Weiss established what would be the prevailing research-based theory on dealing with trauma. "Humans have no conscious access to the non-conscious process that they use to acquire knowledge. Indeed, research into the unconscious acquisition of knowledge demonstrates that the human being has an enormous capacity non-consciously to make inferences."

Weiss described how the nonconscious mind is filled with the raw material of our experiences that influence our reasoning, learning, and decisions, while the conscious mind cannot keep any record of experience. Although numbers differ slightly among studies, the conscious mind can process approximately fifty bits of information per second, while the nonconscious mind can register upward of 200 million bits in the same time (if not more). This information can lodge in the nonconscious mind and is an indelible resource. Every bit of human reasoning, from the fight-or-flight instinct to interpreting interpersonal interactions, comes from experiences registered in the nonconscious mind. Same with physical activity. Just as a baby learns to walk, so we learn the more complicated motions of running, jumping, and catching—through our unconscious mind documenting previous failures and helping us adjust accordingly without conscious thought.

"Most of our learning does not even require that we pay conscious attention. Ninety-nine percent of all learning is a non-conscious process," said Dr. Emanuel Donchin, a specialist in cognitive psychophysiology and former head of the psychology department at the University of Illinois in Urbana-Champaign.

Back on the range, Hebron demonstrated how to train the unconscious. He brought over the group's "least accomplished student, the one with the most to learn." Brandon was a ten-year-old boy with mismatched socks halfway up his calves and a shock of blond hair. Hebron asked Brandon to watch three different swings and say which swing he liked better. Hebron took one swing way outside, one way inside, and one relatively on plane. Brandon, of course, picked the one on plane.

"They know what orthodox is," Hebron said.

He then brought an eight-year-old boy in front of him holding a 7-iron that was up to his chest. Hebron asked the boy if he played baseball. The boy said yes, so Hebron asked him to hit a ball to first base. As the backswing started, the boy took a step back with his right foot—it was a lot for him to swing the long club—then hit one hard to where first base would have been. Hebron then asked him to hit it to third base, and he did. Then he asked him to hit one high to center field, and the boy beautifully flushed one high and straight. "Thanks," Hebron said, and the boy walked away smiling.

"What am I coaching?" Hebron said. "They know what to do. You just have to create the environment to let them do it."


The environment on driving ranges of the PGA Tour event is very different. There, the smallest margin of error separates success and failure. The distinction between better and different is not just semantics—it's a matter of livelihood. For those world-class players, the vast majority of skill development happened long in the past. The depth of their talent for self-direction and self-improvement is staggering. Yet, coaches such as Foley are omnipresent.

"Ultimately, people only know what they know," Foley said. "There is a lot of codependency. [Instructors] out here are making a living; sometimes they tell [players] what they want to hear rather than what they need to hear."

Most instructors, unversed in the scientific process of learning, follow tradition. They dispense technical information because most players crave such technical information. Iain McGilchrist described this modern mind-set in *The Master and His Emissary*, the neuro-anthropological book beloved by Michael Murphy, explaining how the left side of the brain is increasing in power, causing a drastic turn in our evolution as a species. As the left side of the brain gains more influence, we place more emphasis on specific conscious thoughts, the use of language and deductive logic. We place less emphasis on the unconscious mind, where the diminishing right side of the brain functions in images and feelings and general concepts. Given what researchers have discovered about the surprising power of the unconscious mind and its effect on learning, this discrepancy is alarming.

But it is hard to break a habit, especially one that we perceive as necessary. Attempting to consciously amass and absorb information is a left-brain concept, and despite its inefficiency, it has long been the standard of education. The rise of the dominant left brain is the reason why *The Golfing Machine* became such a valuable resource for people who considered themselves swing instructors, and the reason why knowledgeable swing instructors became such valuable resources to players. It is the reason why places such as the Titleist Performance Institute are so popular, and why the information gained from TrackMan is held in such high regard. It is partly why golf course architecture became so unimaginative. It is why players sought out sports psychologists, for a concrete way to solve the problems they thought they couldn't understand.

Yet while our left brain drove us to consciously search for more knowledge about how we perfect our skills, we found that science told us the answers are within.

"The next step in learning golf is realizing the body only moves for one of three reasons: reflexes, anticipation, and when the right side of the brain gives the body a picture to copy and feel," Hebron wrote in his 1990 book, *The Art and Zen of Learning Golf*. Foley read the book closely while he was a young instructor in Canada, and its lessons stuck with him for a long time.

"My suggestion to golfers," Hebron continued, "is to play and learn golf more with the right side of the brain than the left. Pictures, feelings, and concepts are more useful than words when learning any motor skill (golf included)."

Foley, for one, has absorbed the truth in that statement. He saw how Tiger Woods wanted technical information about biomechanics and ball-flight laws when they started working together in 2010, but he also saw how Woods played the game based on pictures and the personal feel he had developed over time. When Foley set out to revive the game of 2016 Masters champion Danny Willet, he did what he could to limit the exchange of technical information despite all of the specifics that were running through his head.

"If you saw a printout of what I was thinking, you'd laugh hysterically," Foley said. "If I get too technical, it may be because I have an insecurity coming through. It's that I want to prove that I'm smart and you should trust me, but it might not be the best thing for the player."

Over Foley's years of teaching, he found that an overload of technical information was often detrimental to a player's development. This phenomenon is known as neurological crowding, as defined by Dr. Peter Huttenlocher, a world-renowned pediatric neurologist from the University of Chicago. "Very ambitious early enrichment and teaching programs may lead to crowding effects and to an early decrease in the size and number of brain regions that are largely unspecified and that may be necessary for creativity in the adolescent and adult," Huttenlocher said, according to the 2003 book *Einstein Never Used Flash Cards.* "It may be no accident that Albert Einstein was a rather average student in his early years [allowing his brain to avoid early crowding effects]."

Yet neurological crowding actually became coveted, as the left brain grew stronger and people wanted to keep feeding it more and more knowledge. The appetite for information, whether useful or not, had become insatiable. In golf, there is no better example of information overload than *The Golfing Machine.* So, of course, that is where Mike Hebron first began.

Two moments changed the way Hebron thought about golf instruction entirely. The first came in 1986, when he was at Smithtown

Landing, waiting for a student to arrive for what Hebron then called "a lesson"—now referred to as "a session," because "the lessons are what the students take away."

Just two years prior, he had published his first book, *See and Feel the Inside, Move the Outside*. This instruction book was like so many before it and so many after it. He had spent years intensely studying *The Golfing Machine*, and that was what he was known for.

"People came from all over because I was a *Golfing Machine* guy," he said. "I knew external clavicle mastoid muscle, I knew all that stuff. And I wouldn't be comfortable standing in front of somebody without knowing that stuff."

He had also picked the brains of local pros as he bounced around assistant jobs on Long Island, following a four-year career as a gritty guard for the basketball team at UNC Wilmington. Gene Borek, the kindly pro at Pine Hollow, told Hebron that he was different from other teaching pros and that he should write his ideas down as much as possible. Hebron started writing every single day in black-and-white marble notebooks, and those notes eventually became the first of his many books. The work was so good that Hebron became the first instructor to have a book stand up as a PGA Master's thesis, earning him the honor of being just the twenty-fourth person in history to hold the title PGA of America Master Professional. The book sold more than one hundred thousand copies, and he was one of the hottest teachers out there.

"Back then, I was in charge," Hebron said. "I had the answers."

As he waited for this student in 1986, he began his normal routine of thinking what he would say. It occurred to him that the student, like many others, had taken the same lesson over and over and was not improving. Hebron was dispensing all this terrific information about the golf swing, and it wasn't helping. Finally, as if a light bulb went on in his head, he realized it was because he knew nothing about the nature of learning.

"I was supposed to be this guru," he said. "I knew my subject, but how do people actually learn? And what was really interesting is, the whole science community had just started studying learning."

The second moment came in 1991, just after Hebron had been given the highest annual honor from the PGA of America, being named National Teacher of the Year. A sales rep from the Mizuno equipment company named Gene McMasters stopped by Smithtown and got to talking with Hebron, who was diving into all there was to know about the nature of learning. McMasters suggested Hebron read *Drawing on the Right Side of the Brain* by Betty Edwards. The bestseller came out in 1979 and would go on to sell more than 2.5 million copies and be translated into at least thirteen languages. The book was partially about learning how to draw pictures, with exercises to help. But, as Edwards wrote, "The true subject is *perception*.

"The larger underlying purpose was always to bring right hemisphere functions into focus and to teach readers how to *see* in new ways, with hopes that they would discover how to transfer perceptual skills to thinking and problem solving. In education, this is called 'transfer of learning,' which has always been regarded as difficult to teach, and often teachers, myself included, hope that it will just happen."

The concept "transfer of learning" hit home with Hebron, and the rest of the book furthered his ideas on the nature of learning. He discovered that trying to put bodily motion into language was hardly helpful in changing a complex motor pattern.

"Language is extremely powerful, and the left hemisphere does not easily share its dominance with its silent partner," Edwards wrote. "The left hemisphere deals with an explicit world, where things are named and counted, where time is kept, and step-by-step plans remove uncertainty from the future. The right hemisphere exists in the moment, in a timeless, implicit world, where things are buried in context, and complicated outlooks are constantly changing. Impatient with the right hemisphere's view of the complex whole, the competitive left hemisphere tends to jump quickly into a task, bringing language to bear, even though it may be unsuited to that particular task."

Using language to teach something such as drawing—or the golf swing—is then unsuitable. Yet by turning away from what so many people wanted, from the left-side language and explicit information that modernity has encouraged, Hebron slowly moved away from extreme popularity. He remained one of the most respected teachers in the game, but became recognized more as a "pro's pro," meaning he was more interesting to other teachers. He gave a lesson once to television personality Charlie Rose, who then invited Hebron on Rose's self-titled, widely viewed interview show on PBS. Hebron started taking many invitations to speak about his ideas and traveled the world, giving talks to PGA pros in such places as Ireland, Italy, France, and Japan. He has spoken to Fortune 100 companies and given lectures at Yale and MIT. Hebron was also the mastermind behind the first PGA Teaching and Coaching Summit, held in Dallas in 1988. Now an annual event, it draws hundreds of pros from all over the world to share their ideas about the game.

But when Hebron came to Orlando in October of 2012 to speak at the World Golf Fitness Summit, a conference that is essentially a promotional and moneymaking event for the Titleist Performance Institute, his message of unconscious learning wasn't exactly embraced—or understood. Most of the audience, professionals from twelve different countries who had all paid a few hundred dollars to attend the summit, became disengaged when Hebron began his slide show.

"When you tell a group of PGA guys that they can't teach anybody," Hebron said, "you lose the room."

In the presentation, he used acronyms, as he often does, to describe some of his thoughts. Up on a big screen in front of a huge auditorium, slides read, "SMART—Students' Minds Are Really Talented"; "SAFE—Students' Always First Environment"; "PLAYFUL—Powerful Learning About Yourself Finds Useful Learning." He ran a video, backed by Queen's "We Are the Champions," with a man with no arms below the elbows swinging a club. Hebron yelled, "Stop!" The video paused at impact, a model standard position, with the handle in front of the clubhead.

"This man shoots between eighty-five and ninety," Hebron said, his voice echoing among creaking chairs and yawns. "He was run over by a car at three years old, and his brothers wouldn't let him make any excuses." Hebron paused to let that sink in. "I don't think golf is a hard game. I think it's inconsistent."

He explained what he thought was one of the biggest lessons any teacher could learn—that emotions are indelibly connected to everything we do, especially our ability to learn. Chemicals—such as dopamine, norepinephrine, and epinephrine—are released in the brain that either help or hinder learning, and their release is a result of the emotion felt while a message is being conveyed. If a student feels scared or threatened, the number of neural pathways (chemicals) available to take in memories and retain information will be limited.

"The way we feel influences how we learn in every learning environment," said Mary Helen Immordino-Yang, a professor at USC and a leading researcher in the field who wrote an influential book titled *Emotions, Learning, and the Brain.* "We attach feelings to things and things to feelings. We always feel something that pushes us on, or away from what we are learning."

As Hebron then told the uninterested audience, "Emotions are the whole deal. You came here—the clothes you have on, the food you eat, how you think—it's based on emotions. When we are working, the brain is taking in new information, emotional information and physical-safety information at the same time. Some children aren't learning at school because they're afraid they're going to get beat up. Some children aren't learning because they're trying to make their parents happy. Some golfers aren't learning because they're trying to make their coach happy. Everything is based on emotions.

"Birds fly, fish swim, people feel. That's all we do."

Foley stood in the back of the room, hanging on some gym equipment. He was still teaching Woods and was the biggest name in instruction. When he eventually got onstage later in the day, he mentioned how Hebron was one of his biggest influences and that it seemed as if no one in the crowd even knew whom Foley was talking about. But when he put up video of Woods and started talking about the technical aspects of the swing, the audience was riveted.

These few hours were like watching a manifestation of our modern longing for language and explicit information, even after Hebron explained to everyone that such a model for teaching was scientifically proven to be inefficient—if not directly harmful—in helping students.

But Hebron did bring up one subject, one name, that created a little buzz. Ben Hogan remained the demigod of golf, the one person who seemed to possess more power over the fickle game than anyone before or after him. Hebron thought of Hogan's famous saying, "Dig it out of the dirt," as an interpretation of self-discovery. And Hebron frequently used one line from Hogan's famous book, *Five Lessons*, to drive home his point.

"If you were teaching a child how to open a door, you wouldn't open the door for him and then describe at length how the door looked when it was open," Hogan wrote in 1957. "No, you would teach him how to turn the doorknob so that he could open the door himself."

"Okay, so you wanted to see the library again, right?" Hebron says in his soft, high-pitched voice. The carpet on the top step leading down to his home basement is neat, picture frames lining the wall going down the left. With about three steps to go, the wall on the right opens up to reveal the small room, piled high with information.

A beige drop ceiling is eight feet high, and the room is bordered and bisected by bookshelves, filled with mostly hardcovers, some aligned upright, some stacked on top of one another. They spill out onto the floor in piles, and no dust is anywhere. The books all look as if they've been opened within the previous week.

The whole length of the left wall is a bookshelf packed with hundreds of plastic-wrapped original editions of *American Golfer* magazine, which stopped publishing in 1936. Above that hang various memorabilia—a framed painting signed by the Big Three: Arnold Palmer, Jack Nicklaus, and Gary Player; a picture of Hebron working with Ian Woosnam at the Masters, which was given to him by friend Hank Haney; a photo of Paul Runyan that Runyan took off this wall and signed during his last visit here.

In a glass case is an original featherie ball from the late 1800s that Hebron bought for a price he'd like to forget. It sits among a handful of Vardon Flyers from the 1910s and 1920s that seem commonplace. A rusted metal gadget from sometime in the early twentieth century looks like a primitive View-Master stereoscope. The gadget has an insert for a slide about the size of an index card that displays 3-D pictures of ancient golfers through what look like old racetrack binoculars, with a long medal handle. Over there is a box filled with a hundreds of these slides.

There is a pull to the middle of this room, where a dark leather golf bag sits atop a small perch, filled with clubs and humming with energy. The bag has a Ryder Cup logo from 1967, and BEN HOGAN is written on the front in white stitching. Looking briefly, one of the irons might have dirt in the grooves, but it's hard to tell.

*That can't be. Not his dirt.*

But Hebron keeps moving, and as the left wall ends, a door hides the unfinished part of the basement, where the boiler rumbles. Keep turning, and then you're face-to-face with a signed photograph, the colors faded.

It's a young Hebron, and he's got his arm around a smiling, bald gentleman. "That's from the first teaching summit," Hebron says. "That was in his office in Fort Worth."

It's the only piece in this extraordinary collection that Hebron seems to take the time to stop to stare at. It's mesmerizing. In it, with his blazer buttoned, the young teacher is almost expressionless.

Just below the picture is another small bookshelf, different from the rest. Instead of displaying shiny dust jackets or faded paper covers, these shelves appear to house old encyclopedias. Get closer, and you see that they're not encyclopedias, but hardcover shells, meant to protect what's inside. Hebron grabs one and flips open the magnetized little door, revealing a first edition of Charles Darwin's *Origin of Species*. Honestly, first edition: London, John Murray, Albemarle Street, 1859.

Almost all the books on this four-foot-high, three-foot-wide unit are first editions. How about a copy of Albert Einstein's *Out of My Later Years*, from 1950, with a pencil signature from the famous scientist himself on the title page? "We play to learn," Einstein writes in chapter 9. Hebron sits cross-legged on the floor, and down at the bottom are the larger books, also in protective casings. *The Art of Golf*, by Sir Walter Simpson from 1887. *The Game of Golf*, by Willie Park from 1896. From that same year, *The Golf Book of East Lothian*, by John Kerr.

"These are the cornerstone books of golf," Hebron says.

Asked for his favorite, he hesitates, then pulls this small red book from a sheath. "*Golfer's Manual*—this is the first instruction book ever written," he says. "1857."

He starts carefully flipping through the rough yellow pages. The book was written under the witty pseudonym Keen Hand, as the author, likely a wealthy nobleman, wanted to remain anonymous. Handing it off to a visitor, Hebron begins to recite the beginning of chapter 4 verbatim:

"'Regarding the practice, no other sport, perhaps on the face of the earth, is there so much difference of opinions as there is in golf.' That's 1857!"

Hebron continues reciting, "'The confusion and multiplicity of styles that prevail amongst players are proof enough of this. Each, no doubt, making his own thinking more correct, but far more superior than any other way.'"

Hebron goes through all his library "often," and has gone to all of this trouble and all of this money for a simple reason. "Maybe you'll learn something from it," he says. "This is where it all started. You know, you have to understand the brain if you're going to understand learning."

Just to the right of this bookshelf, recessed into a stand, is a gray tube television from sometime in the mid-2000s, when the VCR and DVD player built in under the screen seemed groundbreaking. Below it is a handful of VHS tapes, labeled in ink or covered in the case they were sold in. A stack of DVDs rests on the floor.

Around the TV is the sitting area, with a three-person beige cloth couch against one wall, underneath a hanging display of hickory-shafted clubs. Hebron pats the couch lovingly, explaining that it folds out into a bed, one that has slept the likes of David Leadbetter, Dave Pelz, and Runyan. Up on the wall over to the right is a letter written by Harry Vardon, and next to that is a signed picture of Hogan and Byron Nelson at the Masters. Runyan once inspected that picture and started naming the Augusta regulars in the gallery—"Mrs. so-and-so, and Dr. so-and-so."

In front of the couch is a holding rack for more hickory-shafted clubs that have names rather than numbers, maybe fifteen or twenty of them total. Ever pick up an actual cleek? It's like looking down at a cheese knife.

And behind that rises the dark leather golf bag. Now, with some perspective, it's clear that this is the center of gravity. Like a sun at the middle of a solar system, all the things around it are thrown into orbit. The closer you get, the faster you go. The room is a rectangle, and everything—from Darwin to Einstein to Runyan to Nicklaus—is staring toward the middle.

The lights from the ceiling shine off the head of a blond persimmon driver and off the backs of the blade irons, emblazoned with red cursive: *Hogan.*

TEN

# HOGAN

en Hogan was nine years old when he watched his father, Chester, fatally shoot himself in the chest in the front parlor of their house at 305 Hemphill Street in Fort Worth, Texas.

From there, Hogan would spend the rest of his life as a loner, someone content to be by himself and who allowed little insight into his thoughts and feelings. His record as a golfer was only surpassed by the mystique created around his persona.

But why was he revered for his mentality and studied for his technicality? He didn't have the best record and likely didn't have the most talent. So why was Hogan still held as the gold standard in golf excellence?

He turned pro at seventeen years old and fought a nasty hook he had developed trying to win pennies in long-drive competitions as an undersize (and underage) caddie. He almost quit the game a few times and survived on the Tour only by the generosity of loans given to him by a local businessman, Marvin Leonard. Hogan didn't win on the Tour until he was twenty-seven and didn't win a major championship until he was thirty-four, at the 1946 PGA Championship. On Groundhog Day, 1949, he was in a head-on car crash with a Greyhound bus, and doctors told him he would likely never walk again. He only survived because he reached over to protect his wife, Valerie, in the passenger's seat, just missing being impaled by the steering column. Against all odds, he returned to competitive golf a year later and won the 1950 U.S. Open at Merion in one of the most spectacular feats in sports history. He completed the career Grand Slam with his only competitive trip overseas, winning the 1953 Open Championship at Carnoustie in Scotland, returning to a ticker-tape parade in New York City. He spent the rest of his life as a reluctant hero, pouring time and effort into his namesake golf-club company. He died of complications from colon cancer surgery in 1997 at the age of eighty-four, but his legacy lives on in perpetuity—both in the mouths of teachers and in the hearts of everyone who understands and respects the history of the game.

But his epic began back at 305 Hemphill, where a mentality was born and the most worshipped deity in golf history was created.

Almost ninety years later, that ramshackle structure where the most formative moment of Hogan's life took place was a parking lot. Adjacent to what was formerly 305 Hemphill was Hemphill Tax & Notary Service, in a tiny, squat white building with a red pole holding up a slanted roof. It was closed at 1:00 p.m. on a Tuesday. A second small and cracked parking lot separated the tax offices from the Half Price Book Barn, at 321, with beige siding covered with a large sign that read BOOKS in a psychedelic font. Inside sat an elderly woman, Jean, with her white hair perfectly coiffed. She had never heard of Ben Hogan and didn't know of any previous house at 305.

Driving away, around one corner on West Vickery, just south of Interstate 30, was the Fort Worth Recreation Building, a hulking brick warehouse-type structure with letters missing out of its name and a handful of small windows punched out up top. Around another corner was a freestanding lawyers' office, with a car parked out front that was a rolling advertisement—CAR WRECK? CALL 1-800 . . . Around another corner was a sign for a free pregnancy test, then an auto body shop specializing in collision, then a church steeple rising just above the trees.

Neighborhoods change, and history gets lost, but it's hard not to wonder how many people have come around these parts and looked for 305 Hemphill.

There, on the day before Valentine's Day 1922, Chester Hogan put a .38-caliber pistol to his chest and fired a bullet that went right above his heart and out under his shoulder blade. The newspaper accounts differ, but young Ben was almost assuredly in or just outside the room when it happened. His father was "his idol," according to Valerie. Earlier in Ben's life, when the family lived in Dublin, Texas, an old railhead about eighty miles southwest of Fort Worth, Ben used to ease the horses while Chester fitted them for new shoes. But the blacksmith business was essentially eliminated with the introduction of the automobile, and Chester began to suffer from what would now likely be diagnosed as severe bipolar depression, exacerbated by alcohol.

After Chester's suicide, Ben's mother, Clara, moved her family— Ben; his older brother, Royal; and Ben's older sister, Princess—to an apartment on Taylor Street, then to a small house on East Allen. Royal began selling newspapers at the train station, and soon Ben followed. He often had to fight for the best real estate and occasionally slept on the papers he couldn't sell.

When Royal was fourteen, he quit school and started caddying at a golf course called Glen Garden, where he got sixty-five cents for a loop. Ben started following him for the six-mile walk from their house. In the caddie yard, Ben was the youngest and smallest. One time he was put in a barrel and rolled down a hill, and another time he had to run through the paddy wagon as the other boys whipped him with their belts. He was forced to fight an older boy, and when Ben held his own, he was allowed to caddie. The caddies occasionally played with discarded golf clubs, and they used to have long-drive contests on the hardpan of the range. Everyone threw in a penny, winner takes all, and the shortest hitter had to go pick up all the balls.

A caddie named John Byron Nelson Jr. stood out from the rest. He didn't curse, smoke, or drink. He had suffered through typhoid fever when he was young and was tall and skinny and overly polite. He was also the best golfer in the bunch. Eventually, Nelson moved inside to help head pro Ted Longworth in the shop. Longworth's assistant at the time was Jack Grout, who would go on to be Jack Nicklaus's only coach.

On Christmas Day 1927, the day of the annual caddie tournament, Hogan and Nelson had their first epic duel. Nelson made a 40-footer on the final hole to tie Hogan, and they went to a playoff. Hogan made four on the first hole and Nelson made six, but some members wanted to keep it going into a full nine-hole match. Nelson made another tricky putt on the ninth green, and this time it won him the tournament. He was awarded a new mashie (5-iron) and Hogan won a cleek (2-iron). They swapped clubs because Nelson already had a mashie and Hogan already had a cleek. All the caddies were invited inside for turkey dinner, and Hogan was the only one not to attend.

Eighty-five years later, Glen Garden was falling to pieces. It was surrounded by blocks of small ranch houses with chain-link fences, around dusty yards and Virgin Mary statues. Turning onto Glen Garden Drive South, a white-brick sign with green letters declared your arrival. Just to the right of the driveway was another brick wall, with a dirty green-and-white logo, two Gs in a circle divided by crossing

cartoon golf clubs. Behind that wall was the ninth green, slightly raised with a visible back bunker.

The clubhouse, rebuilt in 2000, had a green tin roof and a rock-conglomerate façade. The pro shop was cramped, but the people were cordial, and the dining room was filled with round wood tables and chairs with green plastic cushions. The clatter of pots and pans was constant as the kitchen turned out such dishes as chicken-fried steak with sausage gravy. Near the cart shack was a plain brick memorial with a plaque inscribed with green letters: GLEN GARDEN HONORS ITS FORMER MEMBERS BYRON NELSON, BEN HOGAN, SANDRA PALMER. (Palmer joined the LPGA in 1964 and won nineteen tournaments, including two major championships.) Just past that was the driving range, short on grass. Behind the range, a broken-down fence guarded a yard where a handful of maintenance vehicles were kept, along with piles of blown-out cart tires. The caddie yard used to be there. Now, there was a tall electric pole with a hanging floodlight, listing and close to falling down.

The club's membership had dwindled to almost nothing, and in 2014 the place was sold to F&R Distilling Co., who made decent whiskey.

The most iconic picture of Hogan, and likely the most iconic picture taken in golf history, is the shot of him completing his follow-through with a 1-iron on the seventy-second hole of the 1950 U.S. Open at Merion. The next day, after the eighteen-hole playoff with Lloyd Mangrum that Hogan would win to complete arguably the greatest comeback in sports history, that 1-iron went missing, along with his golf shoes.

Hogan had already become a transcendent star, having previously won a U.S. Open and two PGA Championships. He had dropped out of high school and signed up for the 1930 Texas Open

as a professional, but he withdrew from the first two tournaments he played in, later saying, "I found out the first day I shouldn't even be out there." In early 1938, Hogan told his wife that he would sell his car and clubs if he couldn't make enough money on a five-week West Coast trip. Henry Picard, a venerable older pro, told Hogan that if he needed money, Picard would help him—and that pledge helped relax Hogan enough to finish sixth in Oakland and earn $285 to keep him going. (That, after he woke up on the morning of the first round and saw his car sitting on cinder blocks.) "It was the biggest check I'd ever seen in my life," Hogan said decades later. "I'm quite sure it's the biggest check I'll ever see." He broke through with his first individual win at the 1940 North and South Open at Pinehurst, then quickly became one of the best players on the Tour.

But Hogan became front-page news in 1949 when his Cadillac was crushed by Greyhound bus No. 548, driven by substitute driver Alvin H. Logan, on Highway 80 near Van Horn, Texas. It was Groundhog Day, February 2, which was also the day Chester Hogan would have turned sixty-seven. Ben suffered severe internal bleeding, as well as a fractured shoulder and breaks to his pelvis, left ankle, and a rib, along with several deep cuts and bruises around his left eye that filled it with blood. On the way to the hospital in El Paso, he opened his eyes and asked Valerie where his clubs were. She told him they were in the ambulance with them, safe. He then made a gripping gesture with his hands and mumbled something about hitting a shot toward the gallery on the left. (Always fighting the hook.) When he was in the hospital, he was visited by fellow Tour pro Herman Keiser, who had won the 1946 Masters when Hogan three-putted the seventy-second hole. Hogan looked unconscious until he signaled Keiser over and whispered, "Herman, would you check on my clubs?" On February 18, Hogan needed emergency surgery for a blood clot, and the Associated Press sent out a sixteen-paragraph obituary to hold. He came out of it, but his doctors said he would likely never walk again without assistance. At the

least, the circulation to his legs would be severely restricted, causing him awful pain.

Hogan eventually started pacing laps around his house, then walking out in the street, then playing a little bit of golf—and then he signed up for the 1950 L.A. Open at Riviera. The press coverage and the fan interest was suffocating as Hogan tied Sam Snead for first place and then lost in the playoff a week later, delayed because of rain. It took Hogan hours to get ready for a round of golf—a long soak in a bathtub filled with Epsom salts, followed by a rubdown with Bengay, a single aspirin, and finally the wrapping of his legs in elastic bandages, covered by special rubber-support stockings.

In the famous picture from the U.S. Open later that year, the faint outline of Hogan's leg bandages can be glimpsed through his flowing trousers. But his being able to hit a 1-iron under that much pressure, with that much pain, is what made that photo so enthralling. And it's what made Hogan's club so intriguing to whoever stole it from his bag.

More than three decades later, a collector named Bobby Farino bought a set of MacGregor irons for $150 at a trade show in Ponte Vedra Beach, Florida. After he took his purchase home, he saw that a previously unmentioned 1-iron was included in the deal. The club had a thick cord grip, and the face was hardly worn except for a section the size of a quarter near the hosel. Farino contacted Jack Murdock, a former All-ACC basketball player at Wake Forest. Murdock got his hands on the 1-iron and saw how different it was. He brought it to his pro at Raleigh Country Club, who said he'd never seen anything like it before. So rarely was a club worn-out just in the sweet spot. And a 1-iron? *Could it be?*

In February 1983, Murdock was inducted into the Wake Forest Hall of Fame, and at the induction ceremony he was seated next to PGA Tour player and fellow alumni Lanny Wadkins. He told Wadkins about the curious 1-iron and agreed to ship it to him so he could bring it directly to Hogan, whose equipment company sponsored Wadkins.

"On February twenty-third, 1983, I got a letter from Hogan," Murdock said, before starting to read it: "'Just a note to thank you for making it possible for me to see and possess my old No. 1 iron again. Lanny Wadkins delivered this to my tour director, and he in turn brought it to me. I liken this to the return of an old long-lost friend. Sincerely, Ben Hogan.'"

Hogan eventually sent the club to the USGA Museum in Far Hills, New Jersey, where it was prominently displayed. In 2005, when the U.S. Amateur was played at Merion, the USGA decided to bring the club out of its casing and back to the place where it was made famous. The then executive director, David Fay, was in charge of its transport, and after dinner at a local pub with a Merion member, the two men went out to a park, made sure there were no stones on the ground, and hit a few teed-up balls with the club.


In 1951, a movie was released about Hogan's life called *Follow the Sun*, starring Glenn Ford. By 1953, everyone wanted to know what "secret" Hogan possessed that allowed him to have such control of his game. Rather than disregarding the question, Hogan agreed to be interviewed for an article in a three-year-old magazine called *Golf Digest*. He claimed that the "secret" was a twenty-minute practice routine he did every morning.

"With his feet close together, the Little Man clamps his arms tight against his stomach [and] starts a short swing of a few inches, arms still close," wrote the author, Lawrence Robinson. "Then he gradually lengthens the backswing a foot at a time. He does this for twenty minutes each day until he is taking a full swing, all the while from the close-to-the-body position."

Hogan apparently arranged another deal with CBS to televise his "real secret," but that deal fell through. In 1954, he accepted $10,000 from *Time* magazine for a story titled "Ben Hogan's Secret: A Debate."

Seven pros in the story guessed what it was, with Snead saying, "Any-body can say he's got a secret if he won't tell what it is." On August 8, 1955, *Life* ran the revelatory follow-up "Hogan's Secret." In that article, Hogan told an apocryphal tale about having an enlightening dream in 1946—the year he won his first major—in which the old Scottish term *pronation* came to his mind. Pronation meant that he cupped his left wrist at the top of the swing, making it almost impossible to hit a hook. Combined with a weakened left-hand grip and a "fanning" of the club open on the backswing, there was Hogan's "secret." Really, it was a good way to describe to someone how to slice it.

After publishing his story in *Life*, Hogan began writing his masterpiece, *Five Lessons: The Modern Fundamentals of Golf*, which was serialized in *Sports Illustrated* before being published as a book. Coauthored with renowned golf writer Herbert Warren Wind and illustrated with beautiful pencil drawings from Anthony Ravielli, *Five Lessons* became the bestselling golf instruction book of all time. The public couldn't get enough of Ben Hogan, even if the technical information he shared was harmful to most average golfers—and he knew it.

"I doubt if it will be worth a doggone to the weekend duffer and it will ruin a bad golfer," Hogan wrote at the end of his *Life* article. Many years later, golf historian and writer Al Barkow approached Wind and asked him about the article and its veracity.

"Well," Wind said, "you have to make a living."

Yet the search for Hogan's "true" secret continued.

Jimmy Ballard had his own self-serving theory that went back to the 1945 PGA Championship, when former Yankees outfielder Sam Byrd—Ballard's mentor—made it all the way to the championship match, losing to Nelson in the midst of his eleven straight victories. Hogan was not back from his military service in time to play the 1945 PGA, but when he did get back, he got in touch with Byrd and asked, according to Ballard, "What the hell do you know about putting a stick on a ball that nobody else knows?" Byrd explained Babe Ruth's "connection" philosophy, and Hogan subsequently won his first of nine major championships the next year.

A handful of existing "personal letters" are said to be from Hogan, explaining all sorts of things about the golf swing. Ballard had one such letter and was so secretive about it that he didn't want to share the whole thing. He said Hogan had written the letter to his personal doctor sometime in the 1950s. Ballard said the doctor had come to him for lessons and they made a copy under the strict stipulation that Ballard not share it. In the letter, Hogan had drawn some stick figures, explaining in his swirling and large cursive what he called "a new feel in my golf swing." The language was so candid—referring to his testicle as his "nut"—and so similar to what Byrd taught, that when Ballard first saw it, he balked.

Hogan described the feel: "As the club is taken back, the left knee moves toward the rt [*sic*] knee. . . . At the top of the back swing the groin muscle on the inside of your rt leg near your right nut will tighten. This subtle feeling of tightness there tells you that you have made the correct move back from the ball."

That was the weight shift that Byrd talked about, and the coiling over the back leg that Ruth had mentioned. Ballard felt vindicated. He had his student Dr. William P. Huckin to thank for sharing the letter.

Decades later, Dr. Huckin, a semiretired dentist, told a slightly different tale. While Huckin was a student at SMU in the 1970s, he met Hogan twice. Hogan would be out on the eleventh fairway at Shady Oaks almost every day at 1:10 p.m., hitting balls to a stationary caddie. Huckin had gotten his hands on a four-page document that Hogan used to give his private students—*students!*—at nearby Preston Trails Country Club. Huckin asked Hogan about the stick figures and the lessons, and Hogan confirmed it all.

"I didn't fully understand or appreciate what he was saying," Huckin says. "I really didn't."

In 2017, a Louisiana teaching pro named Larry Miller published a book titled *Ben Hogan's Secret Fundamental: What He Never Told the World.* As a young man, Miller had attempted to play the PGA Tour, and he sought swing advice from Tommy Bolt, the temperamental and respected pro who won the 1958 U.S. Open. Bolt said Hogan had given him the advice to turn his own career around, advice that included Hogan's "real secret." Of course, Hogan told Bolt not to share this secret until he found "a worthy and trusted protégé."

Bolt passed down Hogan's "secret" to Miller, who later published it. Miller's description of Hogan's "geometry" was almost identical to what Hogan had supposedly written and illustrated in Dr. Huckin's letter. Miller also described principles for strengthening the body so that one could swing faster and with more control and also included the final provision that all of the technical directions were to be incorporated with "deep practice," which meant hitting five hundred to six hundred balls per session. That undoubtedly sounded like Hogan.

Barkow, the historian, had another "letter" story, heard from longtime touring pro Loren Roberts. Roberts claimed he had a thirteen-page document that Hogan had written to the head pro at Pasatiempo Golf Club, Pat Mahoney. According to Barkow, Hogan wrote:

"The most important part of a good golf swing is to take the club back correctly so as to keep the head in one place."

That point directly contradicts what Ballard teaches. The letter continues:

"This can be accomplished in only one correct way, by moving the left knee in toward the right knee, while moving the left shoulder in a slight downward arc. . . . It feels like the hips are moving to the right but this is not so."

Near the end of that letter comes some familiar advice:

"To verify a correct backswing, at the top of the backswing the groin muscle on the inside of your right leg near your right nut will tighten. This subtle feel of tightness there tells you that you [can make] the correct move back to the ball."

Barkow also heard Hogan speak firsthand about his "secret." When the PGA Tour started its minor-league affiliate, it was first titled the Hogan Tour, sponsored by Hogan's company. Before the launch of that Tour, Barkow was granted a long sit-down interview with Hogan, during which Hogan said that the real "secret" would be revealed. After much talk over lunch, the two were walking down a hall at Shady Oaks, and Barkow asked Hogan for the secret. Hogan pulled him inside the kitchen so no one else could hear. He told Barkow to set up as if he were going to hit a shot. Then Hogan told him to move his head to the right.

That was it. That was the secret. Moving his head to the right. Barlow asked if it was a gimmick, a trick, a joke—and Hogan swore it wasn't. He said the tip went back to Bobby Jones.

So why, for the following half century after Hogan's success, did teachers use him as an example? Why did so many people scour the achieves looking for ancient video of his swing, hoping to glean the smallest bit of insight into what made him great? Why could no one accept that maybe the answer wasn't a piece of technical information?

Hogan, of course, dug his answers out of the dirt. He symbolized the masculine ideal to be stoic, to not ask for help, to overcome drastic circumstances and odds all by yourself. Golf lent itself to that narrative, people taking extreme pride in it being a game where on-course occurrences represented life lessons. You have to recover from a bad bounce. You have to call a penalty on yourself if you make a mistake. Honor and integrity are paramount. Cheating in any form was worse than blasphemy.

So it was easy for most to believe that Hogan did have a specific and unique secret. One that made his success tangible, something that could be copied and understood if only the secret was revealed. That modern mentality had convinced people there was an answer for everything—you just had to look hard enough. Hogan's intentional ambiguity tempted people to keep searching.

When he was given a ticker-tape parade down the Canyon of Heroes in New York City for winning the 1953 Open Championship and completing the career Grand Slam, a reporter from Reuters asked if he had "perfected" the game. Hogan answered, "I hope I never become perfect. Because then, where do you go from there?" He paused. "Down, that's where."

People's faith in the idea of athletic perfection comes back to that hatred of mystery—the search for inalienable truth, the belief that everything can be figured out over time. It's a point of view that grew in concert with the exponential advances in technology.

But there was always something more to Hogan, something that attracted people more than just the technicality, the Lazarus-like recovery, or that mysterious "secret." Shortly before Bobby Jones died in 1971, he was asked which great player he would select if he needed to win a major tournament.

"That's not hard for me to answer: Hogan," Jones said. "He had the intangible assets—the spiritual."

# OREGON AND STRING THEORY

The eighteenth hole was the silent hole, and I walked along, watching. The sun was inching lower toward the shores of the Pacific Ocean, and the Oregon ground was sandy and firm under my feet. I was following the final group of the Shivas Irons Society, who were playing the last hole of Pacific Dunes, the terrific course designed by Tom Doak at the Bandon Dunes resort. As was their tradition, the Shivas Irons group played one hole per round in complete silence, so I obliged.

Just before a player named Steve teed off, a group of three possums scurried across the fairway. They paused just long enough to jostle one another, look up at the tee, then run away, back into the woods. Hands went up and faces squirmed, but no one said a peep. It almost felt better that way. Finally, with some bad shots and some exaggerated frowns, everyone in the group holed out. About twenty members were waiting behind the green, and they all gave a good-natured cheer when Steve rolled one in for his par.

The October shadows were long, and the air was crisp and starting to cool. It was about 5:00 p.m., and another ninety minutes of light might have been left in the day. Behind the green, Steve spotted Fritz, a tall and square-jawed man. They shook hands and hugged, and Steve introduced me to Fritz, joking that I'd been following him around all day. They laughed, and Fritz told a story about how he finished with an eight on that last hole and it just wasn't sitting right with him.

He looked at me, looked at Steve, and offered a suggestion. "Think we can make it to the Preserve?" Fritz was referring to Bandon Dunes' thirteen-hole par-3 course that plays over some of the most dramatic terrain on the property.

Steve, Fritz, and I got there just in time to hassle the starter before he was gone for the night. On the first tee were eleven moderately drunk men, playing as a single group. That kind of thing, as long as it's good-natured, is allowed at this time of night. The starter said he could get our threesome in behind them, as the last group of the day, "but I'd ask to play through if I were you."

The Preserve was designed by Bill Coore and Ben Crenshaw, and the holes may be short, but they're still wonderfully interesting. The first was playing about one hundred yards, with the pin in the front of the green, just over a false front. Fritz was first to the tee, as he had been waiting more than an hour to try to make a good swing and feel some redemption. He took a sand wedge and swung in an upright and able motion, catching it a little thin. The ball came out low, but not screaming, took one bounce on the front of the green, took another small one, and then—*clank!*—hit the flagstick and fell in.

"Holy shit!" Fritz screamed as he raised his hands and dropped his club. Steve and I doubled over, laughing so hard that no sound came. The group in front of us had already moved on to the next tee, so Steve and I were the only witnesses. By the time we had composed ourselves, Fritz asked us to swear not to mention that it was hit thin. We gave no guarantees.

After the third hole—and after playing through the large group, which had no chance of finishing before sundown—there was a small snack shack. In a back fridge were light-green cans of perfectly chilled Sierra Nevada Pale Ale.

The rest of the round flew by like a dream. Maybe it was the hole in one, maybe it was the aftereffects of spending a glorious day in unimpeded sunshine, or maybe it was the beer. (The course circles back past that shack again, and it was impossible to just walk by.) The sun began to set, and the sky first turned orange, then pink, then a deep and vibrant purple. The moon hung like a scythe. There were good shots and bad—a great knockdown 8-iron over a deep ravine on the 131-yard sixth to within two feet of a back-left pin; a heavy-chunk 9-iron on the 142-yard eleventh that fell into a 100-foot gully of twisted trees and bushes and didn't even reach the iconic pine standing on the other side.

Time was being kind, slowing down just enough not to rush things. Each swing had clarity, and each step of the walk had comfort. It wasn't a race, and it didn't drag. The minutes passed amicably, with our having no desire to hold on or to look forward. We were joined in communal awareness, a group understanding of just being present in the moment. We were trying and not trying, caring and not caring— doing everything and nothing all at once.

The ocean was always in the distance, and the stars began to light up the sky like a planetarium. By the time Steve hit his putter off the tee of the 109-yard thirteenth hole—and got the ball closer to the hole than anyone using a wedge—the sky was dark. We holed out, shook hands, and smiled.

Contrary to Murphy's fiction, there were no floating orbs, no swaths of space occupied by multicolored energy waves. There was no talk about true gravity or allowing for nothingness. There was no talk about the round beyond contented remarks such as "How good was that?"

We all showed up late for the group dinner, cheeks a bit flushed and smiles a bit too big. Over dinner, we talked for a bit about the future of the game, how to attract more people to play, how to enjoy it more. But this was not the table at the McNaughtons' house from *Golf in the Kingdom*, and to my disappointment, there was no night golf, no howling into caves, no unearthing of shillelaghs and featheries.

We were just a group of happy people, sharing a pleasant experience. This, here, was enough.

Some months down the road, I asked Fred Shoemaker, the leader of this bunch, about his attraction to the game, what drew him to it and what kept him engaged. As articulate as Fred could be, he was at a loss for description. Eventually, he came back, reflective.

"I know at some point, golf is going to end for me. It does for everyone. And I'll miss that. I won't necessarily miss a new golf course and all that. But a bucket of balls, nothing but time . . . That's sometimes the ten a.m. sun, or the late afternoon, when it's just right, and everything just seems to fit.

"Somehow, that's when life makes more sense."


Murphy wrote a sequel to *Golf in the Kingdom*. Far less popular than the first book, it's called *The Kingdom of Shivas Irons*. The fictitious Michael Murphy embarks on a new quest, after India, journeying from California to Edinburgh and Moscow in hopes of finding his fabled teacher.

Murphy instead finds a Scottish physicist named Buck Hannigan, who is also looking for Shivas Irons. In a meta-twist, Hannigan has read *Golf in the Kingdom*, and he sees a connection to the work he's doing. "Officially, string theory" is how Hannigan explains his work. "Unofficially, on possible relationships between hyperspace and living systems. Does that make sense? It should."

When I began to ask about his sequel, the real Murphy, the eighty-two-year-old man sitting at a breakfast spot in Northern California on a crisp autumn day, knew what questions I had coming. In his two-book sequence, his deft blurring of fiction and reality has left many trails of exploration wide-open, with some questions intended to be left to the imagination. In this case, he injects modern theoretical physics into this fantastical story. As much as some of the philosophers at the Esalen Institute might claim that all their mystical searching is just as "real" as the experimental work conducted in a lab, it's not. Being able to quantify and record information from an objective standpoint is what separates the two.

But was modern physics on the verge of explaining mysticism? Was Murphy trying to lead people to that assumption without explicitly saying it?

"You're right on it now, right on it!" Murphy said. "As Einstein said, 'The grandest theory stands or falls with empirical disclosure.' But we need theories, even if they're wrong."

In March of 2013, the CERN institute in Switzerland released their findings from an earlier experiment at the Large Hadron Collider, a seventeen-mile tube that runs underground over the Swiss-French border. At the cost of over $10 billion, and over two decades of research, they had finally simulated the energies that were present one- to two-trillionths of a second after the big bang. By slamming protons into each other at extreme speeds they found a particle that was previously alive in theory only, known as the Higgs boson, or the God particle. It begins to explain how matter has mass and allows for the Standard Model of the universe to remain intact.

More than three thousand scientists worked on the project, and they called the results "magnificent" and "beautiful."

Using the Standard Model as a base, string theory—and its grander, more controversial and overarching superstring theory, and its newest, furthest reach, M theory—expands on the possible explanation of the weird universe. Its purpose is to connect quantum mechanics (the nature of subatomic particles, the things that make up atoms) with general relativity (essentially, how gravity works). Scientists have learned that subatomic particles are incredibly strange, and they don't particularly like our established rules of reality. As an example, in "quantum entanglement" two particles become linked so that they act *exactly* the same, *at exactly the same time*, no matter how far apart they are or what is between them—say, the entirety of the universe—and with absolutely no perceivable medium connecting them. Einstein, with all of his articulate wit, called this phenomenon "the spooky action at a distance."

String theory then says that everything in the universe is composed of vibrating "strings" that have no mass. This creates particles, which create atoms, which create the periodic table of elements, which create human beings and everything else in known existence.

It goes even further. The discovery of the fabric of the universe, known as space-time, was the breakthrough of Einstein's relativity theories. Einstein found out that time is relative. As an object moves faster and approaches the speed of light, time goes slower. He discovered that gravity warps space-time. His theories gave humankind a deeper understanding of what surrounds us, and how it affects us. What he found was brilliant, and beautiful.

But through a series of mathematical deductions, string theory proposes that standard four-dimensional reality—the three geometrical axes, plus time—is wrong. String theory proposes a world of ten or eleven dimensions, all of them intertwined in the fabric of everything.

That means we can't even detect at least six dimensions, let alone begin to understand them. If that sounds silly, just think about how everything we can currently measure, all of the planets and stars and dust and energies combined, is presumed to make up only 4 percent of the total mass of the universe. *Four percent!* Scientists estimate that 73 percent of that total mass is "dark energy" and that another 23 percent is "dark matter." There are good theories to what it is and how it acts, but we still can't see it. How ominous.

That is twenty-first-century science. So now what? Now, does it seem so impossible that events or feelings once considered mystical—or criticized as psychosomatic—could actually be part of a reality that we are not even close to understanding?

I had my toes in the water of this scientific exploration when I first met with Henry Ellison long ago in New Jersey. When Ellison spoke about struggling with time, to me, it wasn't an exaggeration or a hyperbolic statement about never knowing the time of the day. It meant he was experiencing some phenomenon once thought mystical that science might be close to explaining. Later, I felt the same thing with that round in Oregon. When I thought about it, I realized I felt it pretty often when I was playing golf. Intellectually understanding that time is relative and actually feeling it are two different things.

This awakening led me to a fine little book about memory titled *Moonwalking with Einstein*, in which the author, Josh Foer, attempts to train himself for the World Memory Championships. "Without time, there would be no need for a memory," Foer wrote. "But without a memory, would there be such a thing as time? I don't mean time in the sense that, say, physicists speak of it: the fourth dimension, the independent variable, the quantity that dilates when you approach the speed of light. I mean psychological time, the tempo at which we experience life's passage." One of Foer's coaches, Ed Cooke, said that he was "working on expanding subjective time so that I feel like I live longer. The idea is to avoid that feeling you have when you get to the end of the year and feel like, Where the hell did that go?"

When that brief round at the Preserve ended, there was no feeling of *Where the hell did that go?* There was no longing or regret. Maybe golf, at its core, is about creating more memories, jamming them into the same four or five hours that would otherwise have seemed to go by faster. Maybe subconsciously, that's what brings us back. Not one good shot, or one good hole, or even the competition. But the feeling that we've extended our time here. After all, our brains have evolved over millennia to help us adapt and survive. Maybe it's only instinctive to want something that makes you feel as if you were living longer.

In a world that we barely understand, who's to say that isn't possible? Who's to say that human beings don't have some control over the constructs that make up reality?

If golf can help you control time, even in the slightest, how can you not play?

# ACKNOWLEDGMENTS

One of the best things about being professionally reared in the newspaper business is you learn "no" is not an answer. The timing of deadlines and the finite space of print have forged many great writers and editors. Saying you have nothing is not an option. That lesson, above all, remains ingrained in me from my time at the *New York Post*. I still believe it has the world's best sports section and is one of the world's great tabloids. It was a dream right out of college: from getting lambasted by sports editor Greg Gallo or pulled off the desk to cover a midday boxing press conference at Gallagher's Steak House by assignment editor Dick Klayman (who wrote the daily budget on a typewriter); to learning horse racing and humor from Vic Cangilosi and Anthony Affrunti; to learning how to clarify language from copy editor Kevin Kenney and stop misusing words from Mike Battaglino; to working late-night agate shifts with such terrifically talented peers in Howie Kussoy and Tim Bontemps.

I've had the extreme privilege to sit next to Larry Brooks for the past seven years, learning more about hockey and how to be a reporter than can be taught at any journalism school. Even more meaningful is that I can call Larry a close friend.

When earnestly starting this book endeavor, my first stop was to one of the true gentleman of the business—and the best general sports columnist in the country: Mike Vaccaro. He is an author of many terrific books, whose talent is only superseded by his generosity of time and spirit. I'll always be indebted for his help. Two other integral early readers were Ted Holmlund and Jonathan Lehman, extremely talented people whose advice was paramount.

I'm also hugely grateful to the current sports editor, the forward-thinking Chris Shaw, who always pushes for everyone to be better, along with deputy sports editor Dave Blezow and assistant sports editor Mark Hale.

To thank all the people that participated in this book would be impossible, as I've essentially been writing it since I played my first round of golf at age 13. Thanks to Tim Higgins and his son, Dan for taking me out that first day. I'll never forget the flush 7-iron I hit that flew over the hill on No. 5. Looking back, that shot changed my life forever. Also changing my life was Jeff Silverman at Villanova, who took an unfocused kid and helped him find his professional passion.

Of all the pro sports, diving into the golf world reveals a group of people who are generally welcoming and willing to give of their time. It's impossible to count the number of interviews I did, but consider me grateful for every one of them. I do have to give special thanks to Sean Foley, who brought me into his confidence and allowed me to see the true scope of this work. I owe you a Heineken (or two).

Many thanks to my agent, Sam Fleishman, who took a chance with a young writer and stuck with the idea for a long, long time.

He finally found supreme editor Jofie Ferrari-Adler at Simon & Schuster, who, along with assistant Carolyn Kelly, has shown super-human patience with a first-time author through countless drafts. It's difficult to overstate the importance of a good editor, and Jofie is the best.

Of course, doing anything like this is impossible without the support of those closest to you. My parents, Gerri and Ray, have always been pillars of stability and love. I have endless gratitude toward my sister, Allie, and her husband, Seamus, whose three children—Regan, Caroline, and Owen—have brought such joy. And thanks to my in-laws, Tommy and Carla, who bring intellect, wit, and warmth to every situation.

Above all, my wife, Claire, has confirmed her saintly status throughout this whole ordeal. Her effort, patience, and kindess are what made this all possible. She keeps my feet on the ground—not an easy task—and is my best friend and my confidante. I love her more every day, and no thank-you can even come close to expressing my appreciation.

BRETT CYRGALIS
*Fall 2019*

# BIBLIOGRAPHY

Ballard, Jimmy, with Brennan Quinn. *How to Perfect Your Golf Swing: Using "Connection" and the Seven Common Denominators.* Trumbull, CT: Golf Digest/Tennis, 1981.

Bamberger, Michael. *Men in Green.* New York: Simon & Schuster, 2015.

———. *To the Linksland.* New York: Penguin, 1992.

Barrett, David. *Making the Masters: Bobby Jones and the Birth of America's Greatest Golf Tournament.* New York: Skyhorse, 2012.

———. *Miracle at Merion: The Inspiring Story of Ben Hogan's Amazing Comeback and Victory at the 1950 U.S. Open.* New York: Skyhorse, 2010.

Benedict, Jeff, and Armen Keteyian. *Tiger Woods.* New York: Simon & Schuster, 2018.

Boomer, Percy. *On Learning Golf: A Valuable Guide to Better Golf.* New York: Knopf, 1946.

Callahan, Tom. *His Father's Son: Earl and Tiger Woods.* New York: Gotham, 2010.

Capps, Gil. *The Magnificent Masters: Jack Nicklaus, Johnny Miller, Tom Weiskopf, and the 1975 Cliffhanger at Augusta.* Boston: De Capo Press, 2014.

Carlucci, Phil. *Long Island Golf.* Charleston, SC: Arcadia Publishing, 2015.

Carse, James P. *Breakfast at the Victory: The Mysticism of Ordinary Experience.* New York: HarperCollins, 1994.

Cook, Kevin. *Tommy's Honor: The Story of Old Tom Morris and Young Tom Morris, Golf's Founding Father and Son.* New York: Gotham, 2007.

Coyne, Tom. *Paper Tiger: An Obsessed Golfer's Quest to Play with the Pros.* New York: Gotham, 2006.

Darwin, Bernard. *Bernard Darwin on Golf.* Edited by Jeff Silverman. Guilford, CT: Lyons Press, 2003.

———. *The Golf Courses of the British Isles.* London: Duckworth, 1910.

Doak, Tom. *The Anatomy of a Golf Course: The Art of Golf Architecture.* New York: Lyons & Burford, 1992.

———. *The Confidential Guide to Golf Courses.* Chelsea, MI: Sleeping Bear Press, 1996.

Dodson, James. *The American Triumvirate: Sam Snead, Byron Nelson, Ben Hogan, and the Modern Age of Golf.* New York: Knopf, 2012.

———. *Ben Hogan: An American Life.* New York: Broadway Books, 2004.

Edwards, Betty. *Drawing on the Right Side of the Brian: A Course in Enhancing Creativity and Artistic Confidence.* New York: Penguin, 1979.

Enhager, Kjell. *Quantum Golf: The Path to Golf Mastery.* New York: Warner Books, 1991.

Eubanks, Steve. *To Win and Die in Dixie: The Birth of the Modern Golf Swing and the Mysterious Death of Its Creator.* New York: Ballantine Books, 2010.

Faldo, Nick, with Robert Philip. *Life Swings: The Autobiography.* London: Headline, 2004.

Farrell, Andy. *Faldo/Norman: The 1996 Masters: A Duel That Defined an Era.* London: Elliott and Thompson, 2014.

Feinstein, John. *A Good Walk Spoiled.* New York: Little, Brown, 1995.

———. *Open: Inside the Ropes at Bethpage Black.* New York: Back Bay Books, 2003.

Finegan, James W. *Blasted Heaths and Blessed Greens: A Golfer's Pilgrimage to the Courses of Scotland.* New York: Simon & Schuster, 2006.

———. *Emerald Fairways and Foam-Flecked Seas: A Golfer's Pilgrimage to the Courses of Ireland.* New York: Simon & Schuster, 1996.

———. *Scotland: Where Golf Is Great.* New York: Artisan, 2006.

Foer, Joshua. *Moonwalking with Einstein: The Art and Science of Remembering Everything.* New York: Penguin, 2011.

Freud, Sigmund. *The Interpretation of Dreams*. Translated by A. A. Brill. New York: Barnes & Nobles Classic, 2005.

Frost, Mark. *The Grand Slam: Bobby Jones, America, and the Story of Golf*. New York: Hyperion, 2004.

———. *The Greatest Game Ever Played: Harry Vardon, Francis Ouimet, and the Birth of Modern Golf*. New York: Hyperion, 2002.

———. *The Match: The Day the Game of Golf Changed Forever*. New York: Hyperion, 2007.

Gee, Darrin. *The Seven Personalities of Golf: Discover Your Inner Golfer to Play Your Best Game*. New York: Stewart, Tabori & Chang, 2008.

Glenz, David, with John Monteleone. *Lowdown from the Lesson Tee: Correcting 40 of Golf's Most Misunderstood Teaching Tips*. Pennington, NJ: Mountain Lion, 2001.

Golf Magazine. *The Best Driving Instruction Book Ever!* Edited by David DeNunzio. New York: Time Home Entertainment, 2012.

Greene, Brian. *The Elegant Universe*. New York: W. W. Norton, 1999.

Gribbin, John. *Get a Grip on Physics*. East Sussex, England: Ivy Press, 1999.

Gummer, Scott. *Homer Kelley's Golfing Machine: The Curious Quest That Solved Golf*. New York: Gotham, 2009.

Haney, Hank. *The Big Miss: My Years Coaching Tiger Woods*. New York: Crown Archetype, 2012.

Haultain, Arnold. *The Mystery of Golf: A brief Account of its Origin, Antiquity & Romance; its Uniqueness; its Curiousness; & its Difficulty; its anatomical, philosophical, and moral Properties; together with diverse Concepts on other Matters to it appertaining*. Sandwich, MA: Chapman Billies, 1997.

Hawking, Stephen. *A Brief History of Time*. New York: Bantam, 1988.

Hayes, Neil, and Brian Murphy. *The Last Putt: 2 Teams, One Dream, and a Freshman Named Tiger*. New York: Houghton Mifflin Harcourt, 2010.

Hebron, Michael. *Play Golf to Learn Golf*. Smithtown, NY: Learning Golf, 2009.

———. *See and Feel the Inside, Move the Outside*. 3rd ed. Smithtown, NY: Learning Golf, 2007.

Herrigel, Eugen. *Zen in the Art of Archery*. New York: Random House, 1953.

Hogan, Ben. *Ben Hogan's Power Golf*. New York: Pocket Books, 1948.

Hogan, Ben, with Herbert Warren Wind. *Five Lessons: The Modern Fundamentals of Golf*. Trumbull, CT: NYT Special Services, 1957.

James, William. *The Essential William James*. Edited by John R. Shook. Amherst, NY: Prometheus Books, 2011.

Jones, Bobby, and Ben Crenshaw. *Classic Instruction by Bobby Jones and Ben Crenshaw*. Edited by Martin Davis. Greenwich, CT: American Golfer, 2007.

Jones, Robert T., and O. B. Keeler. *Down the Fairway*. New York: Milton, Balch, 1927.

Keeler, O. B. *The Bobby Jones Story: The Authorized Biography*. Chicago: Triumph, 1953.

Kelley, Homer. *The Golfing Machine: Its Construction, Operation, and Adjustment: The Star System of G.O.L.F. (Geometrically Oriented Linear Force)*. Seattle: Star System Press, 1969.

Konik, Michael. *In Search of Burningbush: A Story of Golf, Friendship, and the Meaning of Irons*. New York: McGraw-Hill, 2004.

Kripal, Jeffrey J. *Esalen: America and the Religion of No Religion*. Chicago: University of Chicago Press, 2007.

Labbance, Bob, with Brian Siplo. *The Vardon Invasion: Harry's Triumphant 1900 American Tour*. Ann Arbor, MI: Sports Media Group, 2008.

MacKenzie, Alister. *The Spirit of St. Andrews*. New York: Sleeping Bear Press, 1995.

MacRury, Downs, ed. *Golfers on Golf: Witty, Colorful and Profound Quotations on the Game of Golf*. New York: Barnes & Noble Books, 1997.

Marshall, Robert. *The Haunted Major*. London: Grant Richards, 1902.

McGilchrist, Iain. *The Master and His Emissary: The Divided Brain and the Making of the Western World*. New Haven, CT: Yale University Press, 2009.

Merullo, Roland. *Golfing with God: A Novel of Heaven and Earth*. Chapel Hill, NC: Algonquin Books, 2007.

Miller, Larry. *Holographic Golf: Uniting the Mind and Body to Improve Your Game*. Gretna, LA: Pelican Publishing Company, 1993.

Murphy, Michael. *The Future of the Body: Explorations into the Further Evolution of Human Nature*. New York: Putnam, 1992.

———. *Golf in the Kingdom*. New York: Viking, 1972.

———. *The Kingdom of Shivas Irons*. New York: Broadway, 1997.

Nicklaus, Jack, with Ken Bowden. *Golf My Way*. New York: Simon & Schuster, 1974.

———. *My Story*. New York: Simon & Schuster, 1997.

Norman, Greg, with Donald T. Phillips. *The Way of the Shark: Lessons on Golf, Business, and Life*. New York: Atria Books, 2006.

Novosel, John, with John Garrity. *Tour Tempo: Golf's Last Secret Finally Revealed*. New York: Doubleday, 2004.

O'Connor, Ian. *Arnie & Jack: Palmer, Nicklaus, and Golf's Greatest Rivalry*. New York: Houghton Mifflin Harcourt, 2008.

Parent, Dr. Joseph. *Zen Golf: Mastering the Mental Game*. New York: Doubleday, 2002.

Penick, Harvey, with Bud Shrake. *Harvey Penick's Little Red Book: Lessons and Teachings from a Lifetime in Golf*. New York: Simon & Schuster, 1992.

Plimpton, George. *The Bogey Man*. New York: Harper & Row, 1967.

Posnanski, Joe. *The Secret of Golf: The Story of Tom Watson and Jack Nicklaus*. New York: Simon & Schuster, 2015.

Quirin, Dr. William L. *Golf Clubs of the MGA: A Centennial History of Golf in the New York Metropolitan Area*. New York: Golf Magazine Properties, 1997.

Rapoport, Ron. *The Immortal Bobby: Bobby Jones and the Golden Age of Golf*. Hoboken, NJ: John Wiley & Sons, 2005.

Rotella, Dr. Bob, with Bob Cullen. *Golf Is Not a Game of Perfect*. New York: Simon & Schuster, 1995.

———. *The Unstoppable Golfer: Trusting Your Mind & Your Short Game to Achieve Greatness*. New York: Free Press, 2012.

Rubenstein, Lorne. *Moe & Me: Encounters with Moe Norman, Golf's Mysterious Genius*. Toronto: ECW Press, 2012.

Sampson, Curt. *The Eternal Summer: Palmer, Nicklaus, and Hogan in 1960, Golf's Golden Year*. New York: Villard Books, 1992.

———. *Hogan*. Nashville, TN: Rutledge Hill Press, 1996.

Scott, Tim. *Ben Hogan: The Myths Everyone Knows, the Man No One Knew*. Chicago: Triumph, 2013.

Shackelford, Geoff. *The Future of Golf: How Golf Lost Its Way and How to Get It Back*. Seattle: Sasquatch Books, 2005.

Shipnuck, Alan. *Bud, Sweat, & Tees: A Walk on the Wild Side of the PGA Tour*. New York: Simon & Schuster, 2001.

Smiley, Bob. *Follow the Roar: Tailing Tiger for All 604 Holes of His Most Spectacular Season*. New York: HarperCollins, 2008.

Sports Illustrated (Time Inc.). *Tiger Woods: The Making of a Champion*. New York: Simon & Schuster, 1996.

Updike, John. *Golf Dreams: Writings on Golf*. New York: Knopf, 1996.

Valiante, Dr. Gio, and Mike Stachura. *Fearless Golf: Conquering the Mental Game*. New York: Doubleday, 2005.

Wexler, Daniel. *The Missing Links: America's Greatest Lost Courses and Holes*. Chelsea, MI: Sleeping Bear Press, 2000.

Wind, Herbert Warren. *America's Gift to Golf: Herbert Warren Wind on the Masters*. Greenwich, CT: American Golfer, 2011.

———. *Following Through: Herbert Warren Wind on Golf*. New York: Ticknor & Fields, 1985.

———. *The Story of American Golf: From Walter Travis and Francis Ouimet to Arnold Palmer and Jack Nicklaus, a History of the Great Men, Women and Events of American Golf*. 3rd ed. New York: Knopf, 1975.

Woods, Tiger, with editors of *Golf Digest. How I Play Golf*. New York: Grand Central, 2001.

# ABOUT THE AUTHOR

BRETT CYRGALIS is a veteran sportswriter covering hockey and golf at the *New York Post*. He has reported on almost all major sporting events, from postseason baseball to the Stanley Cup Final to the U.S. Opens in golf and tennis. He is an accomplished golfer and a member of the Metropolitan Golf Writers Association, and he lives on Long Island.